Particle Physics
Dark Matter and Dark Energy

David Chapple

Published 2011 by arima publishing

www.arimapublishing.com

ISBN 978 1 84549 477 3

Printed and bound in the United Kingdom

Typeset in Garamond 12/16

abramis is an imprint of arima publishing.

arima publishing
ASK House, Northgate Avenue
Bury St Edmunds, Suffolk IP32 6BB
t: (+44) 01284 700321

www.arimapublishing.com

Front cover image Reproduced by kind permission of CERN (CERN-EX-11465 ©CERN Geneva)

Table of Contents

PREFACE

Particles are the building blocks of the universe, shaping our very existence, assuming of course that the particles are not nebulous quantum objects! For centuries, scientists have sought to discover and understand more about these particles, trying to unlock the secrets of how our Universe was created and what will happen to it in the future.

Thankfully, in recent years, we have discovered quite a lot about what happened in the very early stages of the history of the Universe right through to the present time. But even so, there are many mysteries still to be solved. And, even now, opinions are necessarily changing about how our Universe will meet its final fate. As an introduction to the world of particle physics, this book is aimed at undergraduate level with discussion which covers the range of quarks, leptons and bosons that we know or believe exist and the search for as yet undiscovered particles, including CERN's work in the Large Hadron Collider.

It also considers dark matter, what indicates that it exists and some possible candidates for it, and dark energy, the mystery force that is actually causing the expansion of the universe to accelerate. There are questions at the end of most chapters to provide practice on the topics covered, with answers and solutions provided.

Having initially worked on the design of weapons systems, David Chapple is a physicist with a long career in lecturing in physics and who lectures in the

OUDCE Department of the University of Oxford in particle physics, quantum physics and cosmology.

CHAPTER 1: Atomic Structure

The Ancient Greeks

How did we first get the idea that matter is made from particles?

After all, it doesn't *look* like that. We can't see little dots on solid objects if we screw our eyes up.

Somebody somewhere must first have thought of the idea, and we have to go back some two and a half thousand years to start out story, and we start with the Greek philosophers.

Greek philosophy wasn't based on the scientific method and it didn't seem to be necessary to put ideas to the test as a general rule.

It was sufficient to *reason* out an argument.

Possibly the first idea about matter came from the Greek philosopher Thales of Miletus (624 – 546 B.C.) who postulated the idea that the origin of all matter was water so that water was the only basic element, and he *reasoned* that the Earth was floating in water.

The next move in thought came from Empedocles of Akragas (492 – 440 B.C.).

His idea was a step in the right direction in that he added some elements to the basic idea of water.

He believed that there were four basic elements in the Universe, these being earth, water, air, and fire.

Maybe there is more truth in this than first appears on the surface especially if we substitute the following meanings to the named elements: instead of 'earth' let's say 'solids', and instead of 'water' let's say 'liquid'.

Then instead of 'air' let's substitute 'gases' and finally instead of 'fire' let's say 'plasma'. There we at least have four states of matter!

However, the first real step forward came from the Greek philosopher Leucippus and his student Democritus of Abdera, Thrace (470 – 380 B.C.). Democritus developed the idea that matter was actually made from small indivisible particles, which he named ***'atomos'***, meaning 'indivisible'.

He described these atoms as eternal, invisible, and absolutely small so that their size could not be diminished.

They were also incompressible, without pores, and were homogenous and they only differed in shape, arrangement, position and magnitude.

This was insight indeed given the lack of evidence at the time.

So the scene was set nearly two and a half thousand years ago for progress to be made from what was a great insight.

But the later philosopher Plato (427 – 347 B.C.) thought that the elements were based on 'form' and a further derailment from what was a great insight was encouraged by the great philosopher Aristotle (384 – 322 B.C.) who effectively brought an end to this correct line of reasoning by adopting the view that nothing could be indivisible since that might limit the power of the gods.

However, the quantum world has now enabled us to view this idea in a different light since unobserved particles in the quantum realm can be 'spread out' and appear to be indeterministic.

The fact that the scientific method was not established meant that philosophical reasoning was the accepted way of determining how things were and as a result their conclusions were entirely subjective.

Of course there was no need to investigate anything because all conclusions from philosophising were 'obvious'.

Although Aristotle was a great philosopher, he got some things wrong which could easily have been checked.

For example he believed that material objects sought rest with respect to the cosmic centre which was 'clearly' the Earth and that a heavier object with its greater desire for the cosmic centre would "without doubt" fall faster than a light object.

This could have easily been tested but there was no need since the conclusion was 'obvious' and it took Galileo Galilei (1564 – 1642) to sort this out at Pisa.

Dalton's great insights

John Dalton *(figure 1)*

(Reproduced by kind permission of the Manchester Literary and Philosophical Society)

No major breakthrough was made thereafter from the particle point of view until the great John Dalton (1766 – 1844).

Dalton's insight was profound. He was a chemist who realised that gaseous compounds were made from atoms of different elements and he was able to produce a table of atomic weights.

John Dalton realised that water was a compound of gases and that elements were something different in that every element consisted of 'tiny particles called atoms', and that all the atoms in a given element were identical. He further realised that all the atoms in any given element were different from those of any other element, and that the atoms could be distinguished from one another by their respective relative weights.

In addition he realised that the atoms of one element could combine with atoms of another element to form a compound with each compound having the same relative numbers of different atoms.

His insight at that time was remarkable and he was successful in laying a solid foundation in terms of the likely nature of particles for the great discoveries that were to follow.

The twentieth century proved to be a century full of profound discoveries as far as particle physics was concerned.

This started with a great discovery just before the start of the century – three years before in fact.

J.J. Thomson and the electron

The great physicist J.J. Thomson (1856 – 1940) was interested in what then could have been described as 'mysterious' cathode rays, since nobody seemed to know anything of their true nature.

Prior to 1897 cathode rays had been observed in evacuated glass tubes.

They were named thus since they emanated from the *cathode* which was negatively charged in the thermionic emission end of the cathode ray tube as shown:

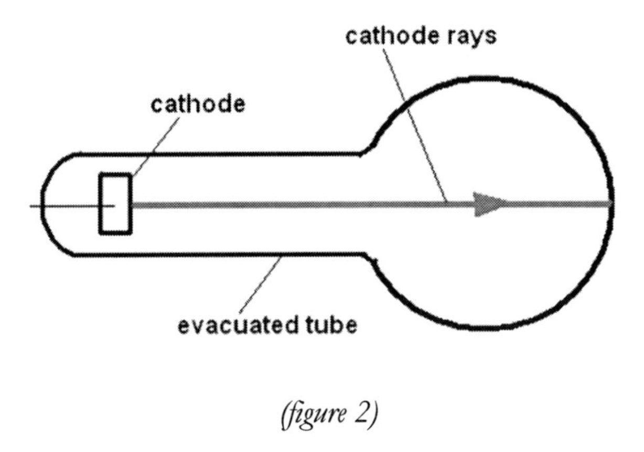

(figure 2)

The cathode rays were detected by a fluorescent coating such as zinc sulphide on the spherical end of the tube.

The cathode rays could reasonably have been either wavelike or particle like in nature.

If they were wavelike they couldn't have been deflected by the superposition of magnetic and electric fields and neither could they have been deflected under these conditions if they had been neutral particles.

However, in 1897 J.J. Thomson succeeded in deflecting the rays by both types of field and this showed that the cathode rays were in fact charged particles.

Furthermore the charge was easily established to be negative due to the observed deflection away from the negative plate in the electric field as shown below:

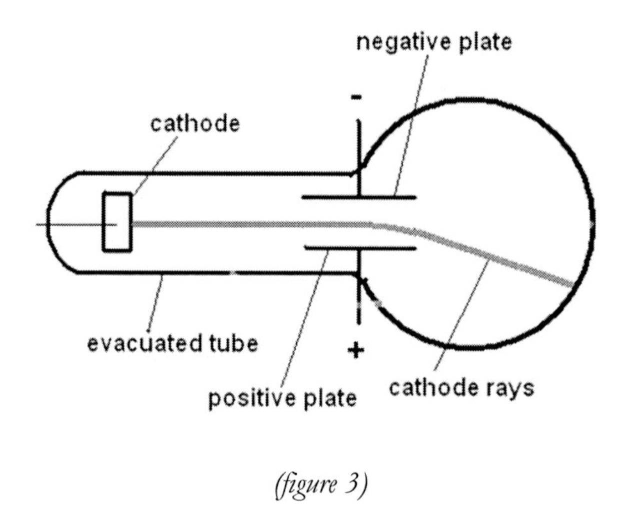

(figure 3)

In fact J.J. Thomson had discovered the electron.

This was really a revolution.

Why? - because at the time the atom was thought to be the smallest particle that existed. In fact the atom was thought to be what we now refer to as a *fundamental particle.*

But since the electron was much smaller than an atom it therefore had to be a *part* of an atom instead.

This obviously meant that the current view of the atom was incomplete.

As a result, in 1904 J.J. Thomson proposed the *plum pudding* model of the atom.

This suggested that the bulk of the mass of an atom was uniformly distributed in the remaining homogenous positive part of the atom, just like the plums in the traditional plum pudding, something like the figure below:

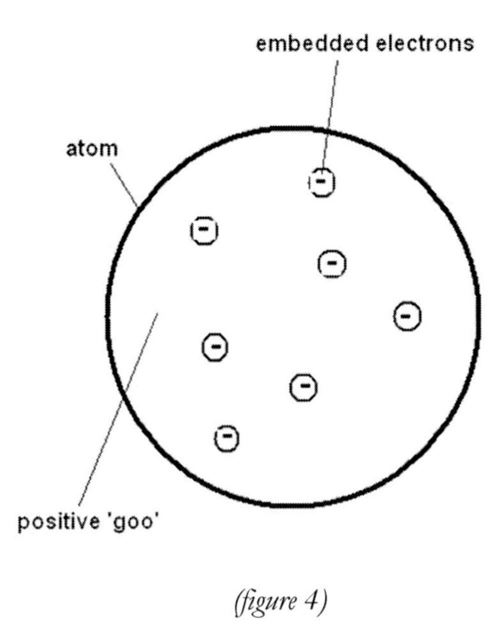

(figure 4)

This model was generally accepted until the ground breaking experiment carried out under the supervision of Ernest Rutherford (1871 – 1937), which was carried out in 1909.

Rutherford's idea was to fire a stream of the quite newly discovered alpha particle at a detector but only after having passed through a very thin film of gold foil, just a few atoms thick.

The alpha particles were expected to pass straight through the gold foil on the centre line when their passage avoided a close encounter with an atom, but some were expected to show some slight deviation on both sides of the centre line.

Why was this?

It was because inevitably some alpha particles would experience a close encounter with an electron which was then known to be negative.

An alpha particle is of course a helium nucleus which is positive and a close encounter with an electron could cause a slight deflection due to opposite charge attraction.

But why was only a *slight* deviation expected?

The answer is that an alpha particle can be thought of as a 'juggernaut' in the particle world, having a mass of some eight thousand times that of the electron, so any deviation would be slight given the huge relative mass of the alpha particles that needed to be shifted.

This was what was expected:

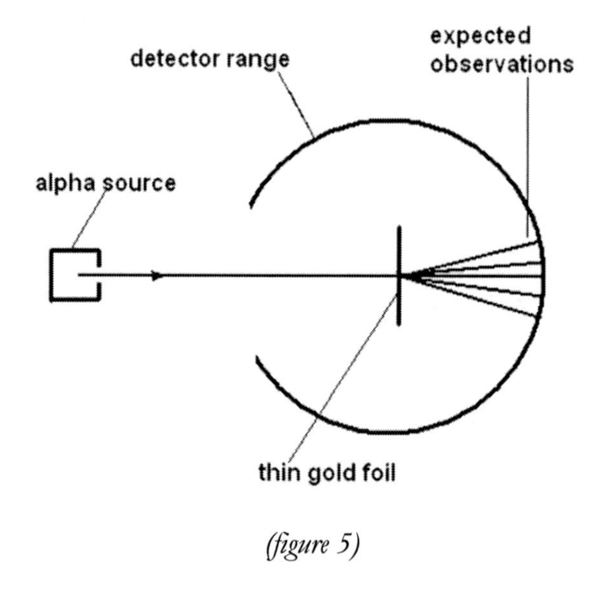

(figure 5)

And then something which was absolutely amazing at the time was seen.

When the detector was moved towards the rear of the gold foil, where no scintillation was expected at all, some scintillations were seen, like this:

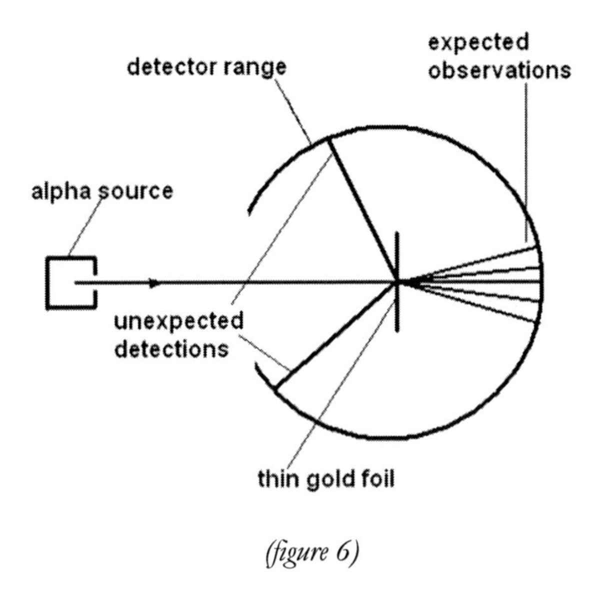

(figure 6)

This was entirely unexpected for the reasons already mentioned but the implications were profound.

The rearward deflections of the alpha particles could be likened to a heavy truck travelling at speed on a road and then seen to 'bounce backwards' without hitting anything!

The only way a heavy truck could bounce backwards would be if it collided with something very massive, probably more massive than itself.

This meant that the idea of the 'positive goo' part of the atom in which the positive charge and bulk of the mass were evenly distributed was wrong.

Clearly the alpha particles must have hit something very massive – in fact much more massive than themselves, in order to deflect with obtuse angles.

In fact they had collided with *gold nuclei* – each one about forty nine times more massive!

Rutherford had succeeded in discovering the next step in the particle story – the discovery of the solid nucleus with a concentrated mass.

This is referred to as the Rutherford model of the atom, as shown in the representation of a helium atom below:

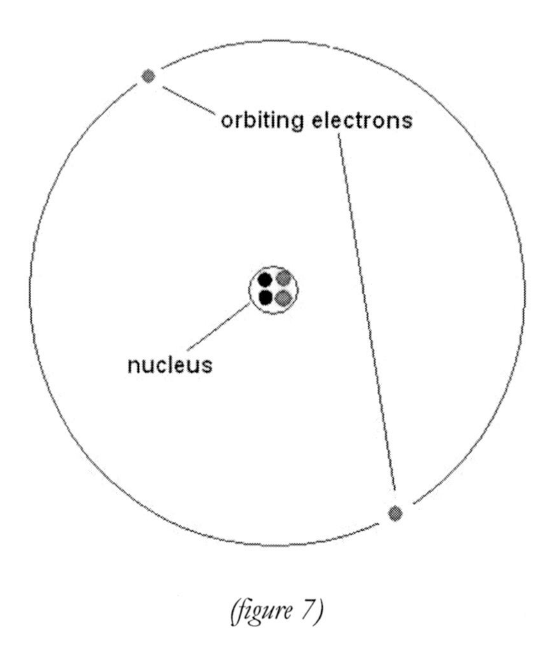

(figure 7)

Since the positive part of the nucleus consists of protons, then effectively Rutherford had discovered the proton.

However, Rutherford had postulated that there was *another* particle besides the proton in the nucleus.

Following on from this, James Chadwick (1891 – 1974) carried out another ground breaking experiment.

Stable beryllium was bombarded with alpha particles and a high mass radiation was observed being emitted from the beryllium.

This radiation possessed sufficient momentum to dislodge protons from a proton rich hydrocarbon like paraffin, but since the radiation was uncharged it was initially thought to be gamma radiation.

The problem was that the radiation would not cause any photoelectric emissions, which gamma radiation was easily capable of doing due to the very high photon energy in the typical order of a million electron volts.

Therefore the radiation could not have been high energy electromagnetic waves, but whatever it was, it was definitely uncharged!

In fact the *neutron* had been discovered, and by momentum conservation Chadwick was able to work out the mass of the neutron.

This is the nuclear equation describing this process:

$$ {}^{4}_{2}H_{e} + {}^{9}_{4}B_{e} \rightarrow {}^{12}_{6}C + {}^{1}_{0}n $$

So by 1932 the constituent particles that make up the atom had been discovered, these being the electron, the proton, and the neutron.

These were all thought to be *fundamental particles*, and they were thought to be the last word.

But they were soon to be proved wrong …..

CHAPTER 2: Fundamental Particles – Leptons and Quarks

Fundamental Particles

The first particle that was thought to be *fundamental* was the atom.

In other words it was thought that the atom could not be broken down into smaller particles – that is, it was thought to be indivisible.

Clearly this was not the case since the atom was then found to consist of various combinations of protons, neutrons, and electrons.

These three particles were also thought to be fundamental but that belief was soon to be overturned.

Leptons

Actually one of the particles in the atom is at this time thought to be fundamental, and this is the electron, which is thought to be indivisible into smaller units.

The electron is a member of a family of particles called *leptons.*

All leptons are thought to be fundamental particles so they cannot be broken down into smaller particles.

What are leptons like?

They can be loosely described as 'lonely' particles.

That is, they don't have a 'social life' in that they never mix with each other and they tend to 'hang out' on their own so that they are always positionally discreet but this is the exact opposite to the other type of fundamental particle that we will meet later.

All the 'main' leptons are charged, and they all have a *negative* charge.

This charge is -1 in atomic charge units, this being 1.60219×10^{-19} Coulomb in magnitude terms.

However, only one of the lepton family is stable in that it does not decay into other leptons.

What is this particle?

1. The Electron, e^-

The most common lepton that we encounter not only on our planet but throughout the universe in a stable state is the electron which doesn't decay into other particles.

How many lepton types are there in total?

There are three charged lepton types, all of them having the atomic charge of -1, and the only difference between them is their *mass*.

As a general rule, the greater mass that a particle has, the more prone it will be to decay into a lighter particle spontaneously in a very short time.

If a particle does decay into a lighter one, it is said to be *unstable*.

In atomic mass terms, the mass of the lightest and most stable lepton, the electron, is zero.

However, in specific terms the electron does have a relatively small mass, this being 9.109565×10^{-31} kg.

Put another way, the electron mass can be expressed as 1/1836.1527 of the proton mass, this putting some perspective on just how small its mass is.

However, particle physicists don't express mass in terms of kilograms nor atomic mass units.

Instead mass is expressed in terms of *energy*.

This is possible since Einstein's energy mass equation states the equivalence of energy and mass.

In these terms, the mass of the electron is given as:

$$\mathbf{m_e = 0.511\ MeV}$$

(Note that sometimes one might see quoted as the electron mass

$$\mathbf{m_e = 0.511\ MeV/c^2}$$

It is important to remember that this is ***not a unit.***

But it is sometimes seen in various texts to differentiate between energy and mass using the **'c^2'** at the end.

However, we will ***not*** be using this notation)

2. The Muon, μ^-

The muon, like all the 'main' leptons, has a charge on -1 in atomic charge units.

But the muon is *unstable.*

This means that it has a short half life until it decays naturally into other particles. How short is this half life?

The muon will only live for some 2.1994 x 10^{-6} second (or just over two millionths of a second).

When its short lifetime is over it will decay into an electron and two other particles, these being an electron antineutrino and a muon neutrino, as described in this well known decay equation:

$$\mu^- \rightarrow e^- + \bar{\nu}_e + \nu_\mu$$

It is worth noting here that the muon and other more massive fundamental particles that decay are not 'made' of the particles into which they decay!!!

If this were the case, they would not be classed as fundamental particles at all.

When these particles decay, it is better to think of them as disappearing from the universe as particles as they convert into pure energy, and from that energy the new particles are formed.

The nature of the neutrino will be described later.

Note now that the solid line centrally above a symbol in equations (or sometimes just below it) denotes the particle's antimatter equivalent.

Because the muon readily decays into other particles, its *mass* must therefore be greater than any of the other particles and in fact it must be greater than the *combined mass* of all the particles into which it decays.

By how much do you think the mass of the muon is greater than that of the electron?

Actually the mass of the muon is two hundred and seven times greater than that of the electron!

So this means that in terms of mass energy, the mass of the muon is

$$\mathbf{m_{\mu} = 105.66\ MeV}$$

Actually muons are created naturally in the Earth's upper atmosphere when cosmic rays interact with atoms of gas there.

Muons that have been created in this manner can be detected at higher levels on the Earth's atmosphere after having descended from the upper atmosphere.

In fact their detection there is a confirmation of the time dilation consequence of Einstein's special relativity theory given that the time taken to transit to the Earth's surface is far greater in our reference frame than the lifetime of a muon!

The muon was first discovered in 1937 and was such a surprise to physicists at the time that a Nobel Prize winning particle physicist declared the well known comment; 'who ordered that?'

3. The Tauon, τ^-

The tauon, like its relatives the electron and the muon, has a charge of -1 in atomic charge units, and since the electron is the only stable 'main' lepton, the tauon, like the muon, is unstable.

This means that it too has a short half life.

How long does a tauon exist for?

Actually its half life is much smaller than that of the muon, and it survives for a mere 2.90×10^{-13} second, this being just about one quarter of a million millionths of a second!

The mass of the tauon is of course greater than that of the electron, but it is more massive than the muon as well. Its mass comes out to be

$$\mathbf{m_\tau = 1777\ MeV}$$

which is some 3477 times the mass of the electron. This also means that its mass is greater than that of a proton!

The tau lepton was discovered in 1975 and its decay modes are well known.

It can decay into a muon, which of course then decays according to the equation we have already seen, or it can decay directly into an electron.

These processes are respectively:

$$\tau^- \rightarrow \mu^- + \bar{\nu}_\mu + \nu_\tau$$

and:

$$\tau^- \rightarrow e^- + \bar{\nu}_e + \nu_\tau$$

I have called these three leptons the 'main' leptons, but besides these there are another three leptons with *zero charge.*

These are the neutrinos, and all neutrinos are stable and have, in the past, been thought of as massless particles.

Billions of these pass through our bodies each second from the nuclear reactions in the core of our star, with no known effect.

However, recent observations as to the quantity of neutrinos picked up by huge underground detectors have indicated that the number of neutrinos arriving on the Earth is far less than would have been expected – unless, that is, neutrinos have a very small mass.

If this is the case as now seems likely, this would enable some neutrinos to oscillate between other types of neutrinos on their journey Earthward, offering an explanation as to why the number of neutrinos observed in detectors seems diminished.

However for our purposes we will regard them as stable.

4. The Electron Neutrino, ν_e

The mass of the electron neutrino is the smallest of all.

What is its mass?

There is a large uncertainty but it is thought to be in the region of

$$m_{\nu e} = 1 \text{ x } 10^{-11} \text{ MeV}$$

Note that this is some fifty one billion times smaller than the mass of an electron.

5. The Muon Neutrino, ν_μ

This is thought to be much heavier than the electron neutrino and its mass may be in the order of

$$m_{\nu\mu} = 0.2 \text{ MeV}$$

6. The Tau Neutrino, ν_μ

The tau neutrino is the heaviest of the three neutrinos and its mass is thought to be about one hundred times that of the muon neutrino, in the order of

$$m_{\nu\mu} = 20 \text{ MeV}$$

Particle Generations

Note that all the 'main' leptons only differ in their mass, as do all the lepton neutrinos, and when otherwise identical particles differ because of their mass alone, they are said to be in different *generations.*

This means that the electron, the muon, and the tauon are different generations of lepton particle, and this is also true of the neutrinos. The six leptons in their generations can be summarised like this:

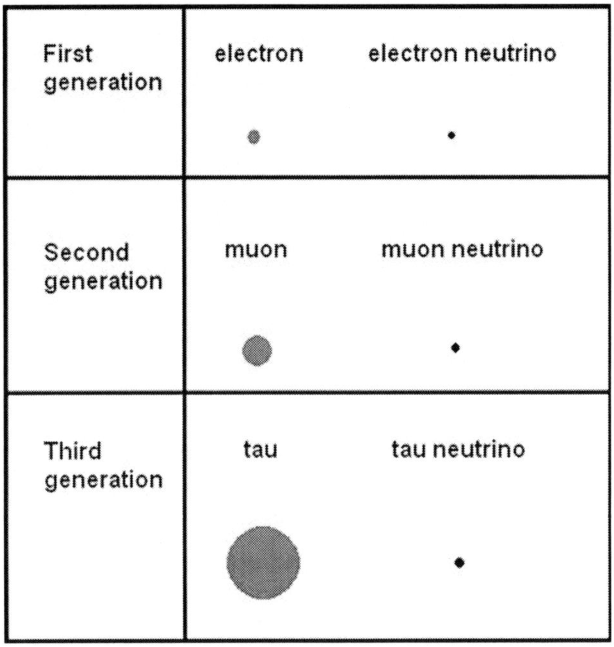

First generation	electron	electron neutrino
Second generation	muon	muon neutrino
Third generation	tau	tau neutrino

(figure 8)

So can you now say which is the only lepton 'in every day use'?

It is of course the stable electron.

There are in the order of 1×10^{28} of them in our bodies (or ten billion billion billion) and they are in all the baryonic matter in the whole of the universe.

Quarks

Murray Gell-Mann *(figure 9)*
(Reproduced by kind permission of Murray Gell-Mann)

In 1961 the physicist Murray Gell-Mann postulated the idea that the proton and the neutron were not fundamental particles after all.

He argued that there might be constituent particles making up protons and neutrons. After all, the atom was once thought to be a fundamental particle but that turned out to be untrue.

The constituent particle that was proposed to make up protons and neutrons were called *quarks*, and at this time quarks are thought to be fundamental particles just like leptons.

What is the nature of quarks?

They can be thought of as 'opposite' in nature to the leptons for the following reasons.

Firstly they are 'sociable' particles in that, unlike the leptons, they are always found in close groups and in fact they cannot exist alone, at least not at the temperatures we take as 'normal' on our planet and in the Universe in general, although things would have been different in the extreme conditions just after the Big Bang.

Further, they all have a type of electric charge, referred to as *colour charge*, but the charges on different quarks are not necessarily the same.

Each quark has a different name which is referred to as a *flavour*.

One way to remember which type of quark has which sort of charge is by their style of name – quarks with what can be taken to be a 'nice' flavour are always positively charged, while quarks with a 'nasty' flavour are always negative, as we will now see!

1. The Up Quark

We start off by taking the word 'up' as being 'nice'!

This means that up quarks are positively charged, and in atomic charge units it has a charge of $+2/3$. All quarks have mass, so what is the mass of an up quark?

Its mass is in the order of

$$m = 4 \text{ MeV}$$

Up quarks are stable and they don't decay spontaneously into other quark types.

2. The Down Quark

Now since 'down' is a 'nastier' name than 'up', then it follows that a down quark must be negative. In fact it has a charge of -⅓

Down quarks are more massive than up quarks and weigh in at about twice the mass of an up quark, so that the down quark mass is given as in the order of:

$$\mathbf{m = 8\ MeV}$$

In most cases, down quarks are stable, although, as we shall see below, there are circumstances where a down quark can decay.

All the atoms in our bodies contain down quarks and they stay that way for millions of years, but there are certain circumstances where a down quark will spontaneously decay into an up quark.

Remember that it can do this because it has a greater mass than an up quark.

A down quark may decay into an up if it is:

1. A constituent of some unstable radioactive isotopes
2. And also in the case of a non-embedded neutron.

Just as the leptons could be segregated into particle generations, quarks can too, and the up and down quarks constitute the first quark generation.

3. The Strange Quark

What sort of name do you think this is – 'nice' or 'nasty'?

We will take it as nasty in comparison with its opposite number which we will meet shortly.

Since strange quarks have a 'nasty' name they are negative and what is more, all the negative quarks have the same charge of -⅓, in atomic charge units.

As far as mass is concerned, the strange quark is heavier again and it weighs in as follows:

m = 150 MeV

4. The Charm Quark

Since the 'partner' of the strange quark, the charm quark, has a 'nicer' name, it will be positive.

Just as all the nastier named quarks have the same negative charge, so the nicer named quarks have the same positive charge, and this is of course +⅔, again in atomic charge terms.

As far as mass is concerned the charm quark is some ten times more massive as its 'partner', and its mass is therefore some

m = 1500 MeV

The charm and strange quarks are second generation quarks and in general they are unstable, with the more massive charm quark readily decaying into a strange quark.

5. The Bottom Quark

Once again the bottom quark can be thought of as having a 'nasty' name, and is therefore negative with the same charge of -⅓.

Its mass is greater again than that of the charm quark and it is

m = 4700 MeV

6. The Top Quark

Our final 'main' quark has a 'nice' name so it has the common positive quark charge of $+\frac{2}{3}$.

What is the mass of a top quark?

Actually its mass is out of all proportion to what we have seen so far and it outstrips its fellows by a huge margin, its mass being a whopping

$$m = 176{,}000 \text{ MeV}$$

Note that this is some three hundred and forty four million times the mass of an electron and approximately one hundred and ninety times the mass of a proton.

As far as particles go, it is the behemoth!

The top quark is unstable and will readily decay into its partner the bottom quark.

The top and bottom quarks are third generation quarks.

Notice that the only stable quark is the up quark, all others being unstable.

The Quark Generations

The six quarks in their generations can be summarised like this:

First generation	down	up
Second generation	strange	charm
Third generation	bottom	top

(figure 10)

Fermions

All the particles that we have considered are called *fermions*, and a fermion is defined as having *half integer spin* such as 1/2, 3/2, etc.

All quarks and leptons have spin one half.

There is another type of particle that has *integer spin* and we will encounter that type of particle later.

Notice that there are not only six leptons, but also six quarks so that symmetry is the order of the day.

However, that is not the end of the story, because we can view leptons and quarks as having not six, but *twelve* members each.

How can this be?

The answer is that for every lepton and quark there can exist an *antimatter doppelganger*.

This means that every particle can have its equivalent in antimatter, such that the antimatter version of the particle is thought to be identical in every way except one – and that is its charge.

Each antimatter particle will have the opposite charge to its matter doppelganger, and should a matter particle come into contact with its antimatter equivalent, they would both annihilate immediately with their mass equivalent *energy* taking the place of the mass.

Remember that all the leptons and quarks that we have met are fundamental particles, but these are just the basic building blocks of the baryonic matter part of the universe. This means that we can build up a vast array of larger particles from these fundamental ones – these are the non-fundamental particles .....

Question

(a) Are stable particles somehow 'inside' the more massive unstable particles?

(b) Describe how stable particles appear from the decay of more massive unstable particles

CHAPTER 3: Mass Energy Conversion

We have already seen that more massive particles can decay into lighter ones.

What exactly is going on here?

The answer was provided to us by Einstein in his great discovery that mass and energy are interconvertible.

That means that all matter can be converted into pure energy.

Don't get this confused with an energy conversion that might be more familiar to us, such as a car's fuel supply for example.

Certainly energy is converted there.

A tank full of fuel can provide kinetic energy to a car and sure enough after many miles the tank is empty, but that isn't what is being described here.

As we know, most of the mass of the original fuel in the tank is still here, appearing as combustion products in the atmosphere such as particulate pollutants and carbon dioxide.

But Einstein's mass energy equation means that when mass is converted into energy there is absolutely none of that mass left – it has all 'disappeared' and been converted into pure energy.

The amount of energy that can in theory be claimed from the perfect conversion of mass seems huge. We can calculate that amount of energy from Einstein's mass energy equation:

$$E = m c^2$$

Now since the speed of light in vacuo, 'c' is a very large number it follows that the amount of energy obtained seems large because the square of 'c' is involved.

The speed of light, 'c', is 2.997924591×10^8 ms^{-1} and if we could convert 1 kg of mass into energy we would end up approximately with a whopping one hundred million billion joules of energy and zero mass!

Trinitrotoluene (TNT) is a very powerful explosive and 16,000 tonnes of it would cause immense devastation if it exploded.

However, a nuclear bomb with a similar power might only contain something like 60 kg of a chain reaction capable isotope, and of that only a mere 600 mg of its mass might be converted into pure energy.

Remember that this is much less than one gram being only about 0.0006 kg.

Compare this with 16,000 tonnes, which is 16,000,000 kg!

Let's get some perspective on this.

Imagine that some time in the future power stations were able to provide power by nuclear fusion, and let's speculate further that some 300g of matter are converted into electrical energy.

How much energy is that?

$$E = m c^2$$

$$E = 0.300 \times (2.998 \times 10^8)^2$$

$$E = 2.696 \times 10^{16} \text{ J}$$

A 1 kW fan heater will consume 3.6 million Joules of energy in an hour, so how long would that three hundred grams worth of energy keep it running?

If you worked it out you should have got 7.49×10^9 hours, or about 855,000 years!

As we know we don't have that facility to produce power yet, but something similar but on a much smaller scale happens daily here on Earth – in the upper atmosphere to be precise.

Here cosmic ray particles collide with and interact with gas atoms and can annihilate into pure energy from which an array of new particles are created, some of which can actually be detected on high ground on the Earth's surface.

Like this…

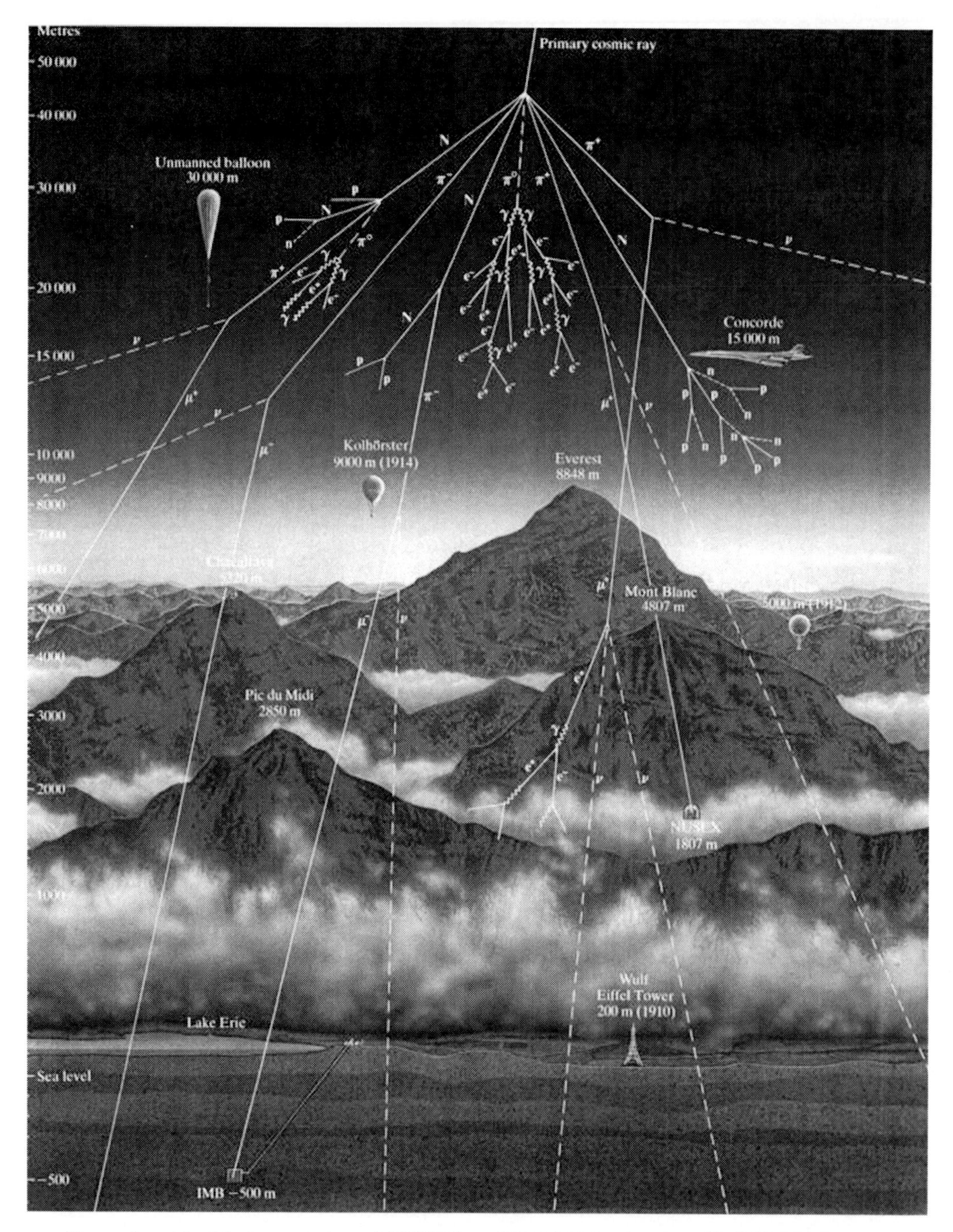

(Reproduced by kind permission of CERN – CERN-DI-9905005 ©CERN Geneva)
(figure 11)

We can categorise a particle's mass in two ways; a particle can have a rest mass and it can also have relativistic mass.

Not all particles have rest mass.

For example, an electromagnetic photon has no rest mass at all.

And yet such a particle does possess momentum!

It was thanks to Albert Einstein that we can view electromagnetic photons as particles, although the experimental confirmation that photons possess momentum was left to A. Compton who received the Nobel Prize for this in 1927.

If we consider photons to be particles, then they have no rest mass, and only particles without rest mass are permitted to travel at light speed.

No particle with rest mass can travel at light speed, and this is due to the huge increase in its mass as it approaches light speed.

However, particles with rest mass can reach speeds *approaching* light speed.

So photons, which have no rest mass at all, possess momentum, and this can be interpreted as a possession of mass at high speed, this mass being relativistic mass.

We however are only interested in the conversion of *rest mass* into energy and vice versa.

To do this effectively we must use the correct units.

It would be completely unrealistic to measure the diameter of a human hair in kilometres or miles!

Likewise, the masses of subatomic particles are so small that is unrealistic to measure them in tonnes!

To solve this problem we use units of mass and energy that are more appropriate and energy is measured in this case in *electron volts*.

First let's define the electron volt (eV):

Electrons are repelled by a negative polarity and attracted to a positive one, so the electron in the following diagram will be attracted to the positive plate and will move over to it as long as it is ionised:

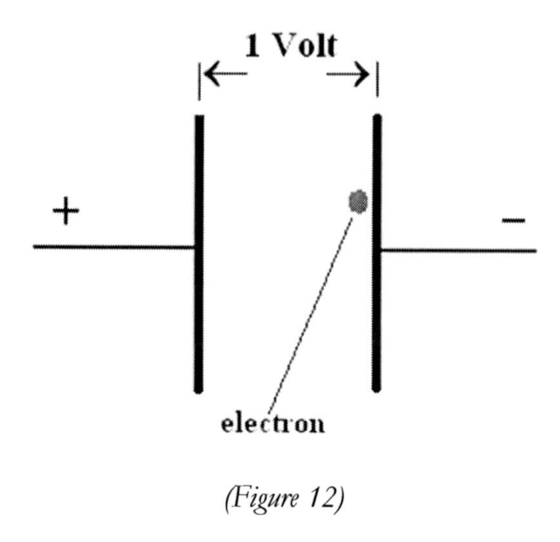

(Figure 12)

When the electron arrives at the positive terminal, work will have been done:

$$\text{Work done} = QV$$

Since the electronic charge is 1.602193×10^{-19} C and the potential difference through which the electron has been accelerated is 1V, then

$$\text{Work done} = (1.602193 \times 10^{-19}) \times 1$$
$$\text{Work done} = 1.602193 \times 10^{-19} \text{ Joule}$$

This amount of work is equal to one electron volt of energy, so the definition of the electron volt is the amount of energy involved when an electronic charge is moved through a potential difference of one volt.

By way of example, let us now calculate the amount of energy that is bound up in a proton, the rest mass of which is 1.672614×10^{-27} kg:

Taking the speed of light in vacuo, 'c', to be 2.997924×10^{8} ms^{-1};

$$E = m\,c^2$$

$$E = (1.672614 \times 10^{-27}) \times (2.997924 \times 10^{8})^2$$

$$E = 1.508270 \times 10^{-10}\ J$$

This must now be converted into the appropriate unit of energy which is the electron volt, and since there are 1.602193×10^{-19} Joule in 1 eV, then

$$E = \frac{1.508270 \times 10^{-10}}{1.602193 \times 10^{-19}}$$

$$E = 9.38257 \times 10^{8}\ eV$$

$$E = 938.257\ MeV$$

In chapter 1 we noticed that the mass of particles is usually expressed in terms of energy, and the result above shows us how this is achieved.

The *mass* of a proton is therefore expressed by the same number. Hence the mass of a proton is expressed as:

$$\mathbf{m_p = 938.257\ MeV}$$

If you are puzzled as to how it is possible to describe energy and mass with the same quantity and units, think about the electronic charge of

1.602193×10^{-19} Coulomb being expressed as 1.602193×10^{-19} Joule as the equivalent energy of 1eV!

Photons

At this point let us consider the nature of light for a moment.

What are photons?

It was in the seventeenth century that Isaac Newton passed light through a prism and discovered that it became separated into the 'seven colours of the rainbow', namely red, orange, yellow, green, blue, indigo and violet.

This, of course, was caused by the refraction of the different wavelengths of light separating out in the glass.

But Newton thought that light was a stream of particles, which he called 'corpuscles'.

Around about the same time there were others that thought differently, namely Christiaan Huygens, who thought that light was a wave, but generally Newton's idea prevailed.

However, it wasn't until the early nineteenth century that the great Thomas Young performed the double slit experiment, which showed that light interfered, thus proving that light was wavelike in nature.

However, in the early twentieth century, Newton was partially vindicated!

It was Albert Einstein who showed that, in fact, light was particle like in nature!

He showed that light was quantised and that it travelled in the form of photons.

A photon can be regarded as a 'minimum packet of energy'.

Light is dualistic in nature.

If we want to show that light is wavelike, it will readily interfere in a two slit type of experiment (as long as it doesn't 'suspect' that it is being observed).

However, if we wish, light will oblige us by showing us that it also possesses momentum (more of that later), thus behaving as a particle.

Question

Data: **Speed of e.m. radiation in vacuo, 'c' = 2.997924×10^{8} ms^{-1}**

Number of joules in an electron volt = 1.60219×10^{-19}

Calculate the energy bound up in the rest mass of the following subatomic particles and thereby quote their mass in MeV

(a) the electron of rest mass 9.10957×10^{-31} kg

(b) the neutron of rest mass 1.674920×10^{-27} kg

CHAPTER 4: Combination Particles – Hadrons

Combination Particles

We have seen how both leptons and quarks can be grouped in generations.

Many different particles can be built from the fundamental particles and these can be called *combination* particles since they are non-fundamental and consist of various combinations of the leptons and quarks.

It is worth noting that *all* the known *stable* particles in the universe are built from the first generation leptons and quarks.

Now it follows that additional combination particles can also be built from all the second generation leptons and quarks as also can other combination particles from the third generation leptons and quarks.

But it must be remembered that all second generation combination particles and all third generation combination particles are *unstable.*

These can be referred to as *exotic* particles and if they are created they will decay into more stable particles.

Second and third generation particles do not make up the stable matter which we see around us – these having all been made from the first generation leptons and quarks.

We can now classify the leptons and quarks together in their respective generations like this:

First generation	down	up	electron	electron neutrino
Second generation	strange	charm	muon	muon neutrino
Third generation	bottom	top	tau	tau neutrino

(figure 13)

Hadrons

Many of the larger non-fundamental particles that are built from quarks are called *hadrons.*

There are no leptons in hadrons, and their only constituent particles are quarks and sometimes antiquarks.

Hadrons can be thought of as two non-fundamental particle types, these being *baryons* and *mesons.*

1. Baryons

Baryons are always made up from three quarks, and the most common baryons by far are two particles with which we are very familiar – these being the proton and the neutron.

Protons are composed of two up quarks and one down quark, something like this:

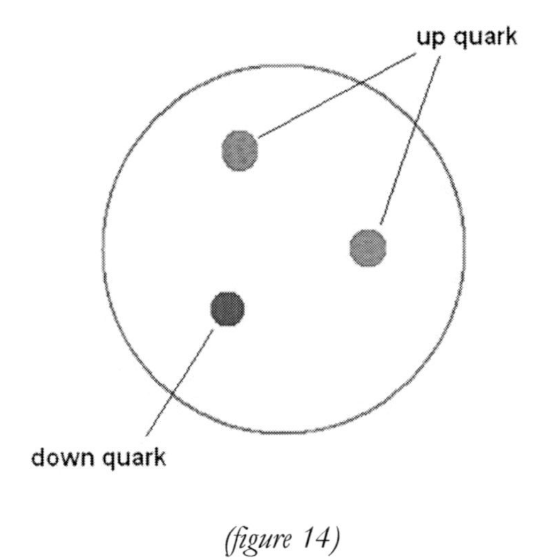

(figure 14)

Can you now work out the charge on a proton from your knowledge of the quark fractional colour charges before you look at the answer below?

Since a proton consists of two up quarks and a down quark then the total charge must be **+⅔** added to **+⅔** which is then added to **-⅓**. This of course comes to a total charge of **+1** which is the charge on a proton.

The rest mass of a proton is:

$$m_p = 938.2595 \text{ MeV}$$

Neutrons on the other hand consist of two down quarks and one up quark, something like this:

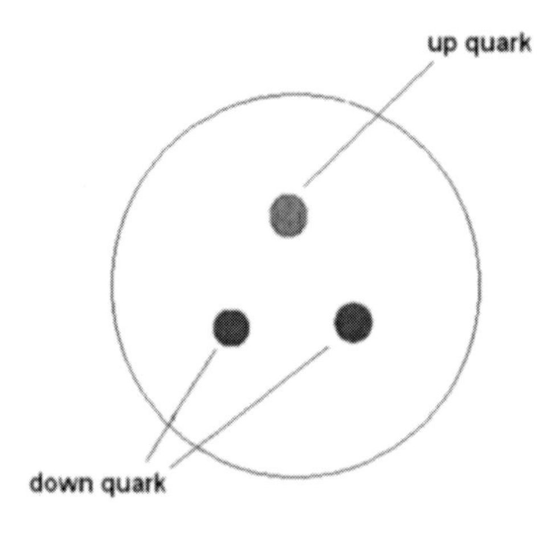

(figure 15)

Can you now work out the charge on a neutron using your knowledge of the fractional quark charges?

You should have found this to be neutral, since two charges of **-⅓** must be added to a single charge of **+⅔** giving zero.

The rest mass of a neutron is slightly more than that of a proton. It is:

$$\mathbf{m_n = 939.55355\ \ MeV}$$

Events involving particles

When something happens to a particle or more than one particle, this can be called an event.

We will consider two types of particle event, namely *decays* and *reactions.*

The decay of a particle is an event which will take place naturally without any external input being involved. Decays are common place in particles that have a relatively large mass.

But reactions involve additional particles as well.

In a particle accelerator, particles are made to collide into other particles, at great speed.

This is usually approaching the speed of light.

Sometimes these particles are in a stationary target and sometimes they are also moving towards the incoming particle, but when they collide there are new particles formed and some of these will then decay naturally.

Now protons are extremely stable and no natural proton decay has ever been observed, and the half life of a free proton has been calculated to be of the order of 1×10^{35} years, or a hundred million billion billion billion years!

However it's a different story for the neutron.

Certainly embedded neutrons in the nuclei of *stable* atoms do not decay, but in the case of a free neutron, decay would happen in about 880 seconds.

Can you now think of what *must* be one difference between a proton and a neutron for this to happen?

I hope you realised that a neutron must be *more massive* than a proton for this to happen.

In fact, the decay of a neutron, whether free or embedded in an *unstable* nucleus, will result in the production of a proton according to the well known beta decay equation:

$$^{1}_{0}\mathrm{n} \rightarrow {}^{1}_{1}\mathrm{p} + {}^{0}_{-1}\beta + \bar{\nu}$$

This describes the neutron decaying into a proton plus an electron in the form of beta radiation plus an electron antineutrino.

Remember that the line centrally above a symbol in an equation denotes the anti particle equivalent of the particle.

Since the mass of a neutron is slightly more than that of a proton then the mass equivalent energy between the neutron and the proton that is formed has appeared in the beta particle and the antineutrino, plus whatever energy these particles now possess, which is mainly in the form of kinetic energy.

Since the neutron decays into the proton, what has happened to the quark structure?

The answer is that one of the down quarks in the neutron has changed into an up quark so that the baryon is now a proton.

But remember that the converse cannot spontaneously occur, i.e. a proton cannot *spontaneously* change into a neutron since the proton mass is less than the neutron mass.

However, it is possible to effect the transformation of a proton into a neutron, but only by the *addition of energy.*

The proton and the neutron are the only stable baryons, but as I said before there are an array of exotic and unstable baryons that can be created and which can exist for a short life span.

One example is the *neutral lambda* baryon which can be created in a particle accelerator and which only exists for some 2.51×10^{-10} second, or just over two ten billionths of a second!

This particle consists of three quarks, namely one up quark, one down quark, and one strange quark, like this:

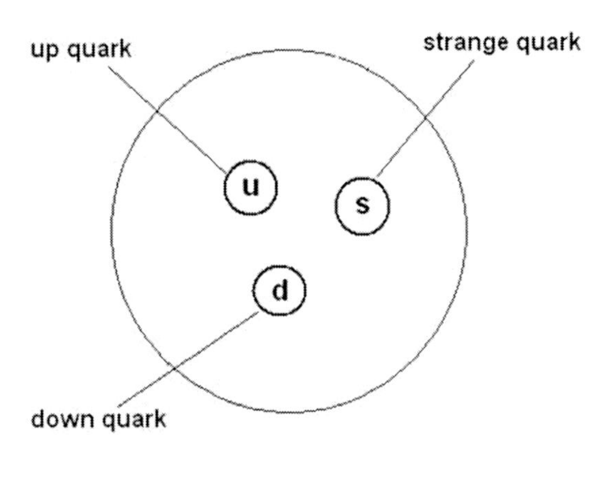

(figure 16)

Now use your knowledge of the fractional quark charges to confirm that this particle has zero charge.

2. Mesons

Just like baryons, mesons are hadrons, but unlike baryons with their three quarks, mesons only have two quarks, and that last statement is a little ambiguous in that one of them is always an *antiquark*.

This means that *all* mesons are composed of a quark and an antiquark.

Can you immediately see a problem with that?

The problem that you might have seen is that a quark and its antiquark doppelganger will annihilate when in contact, so you may think that such a meson couldn't exist. This is true in most cases, so that the quark in a meson is usually accompanied by an antiquark of a different flavour, so that they don't annihilate.

An example of one is the positive K^0 meson, or neutral kaon. This meson consists of a down quark and an anti strange quark, as shown below:

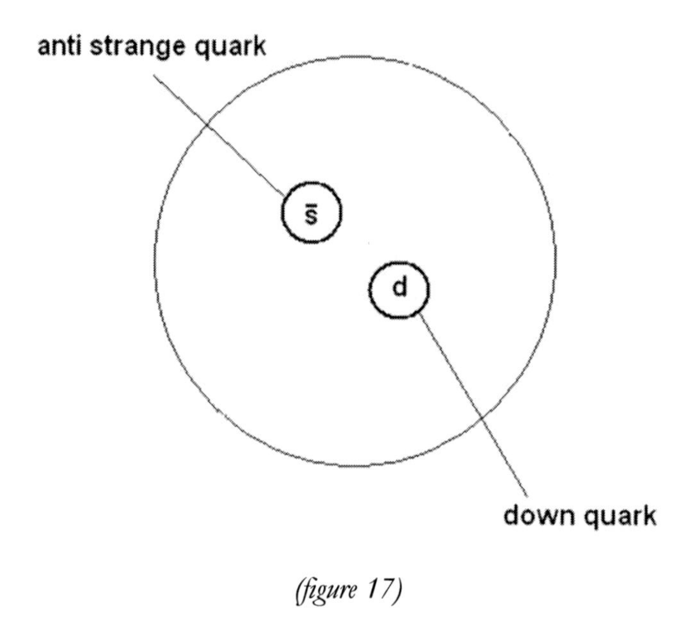

(figure 17)

Using your knowledge again of the quark fractional charges, check that this meson is in fact neutral, but remember that the charge on an antiparticle is always the *opposite* to that of its 'normal' matter counterpart. For example, the charge on an up quark is, as we know, $+\frac{2}{3}$, so it follows that the charge on its antimatter equivalent, the anti up quark must be $-\frac{2}{3}$.

However, mesons with a quark antiquark match of the *same* flavour can be produced but these have an extremely short half life. An example of one of these is the neutral pi meson, which consists of an up quark and an anti up quark as shown below:

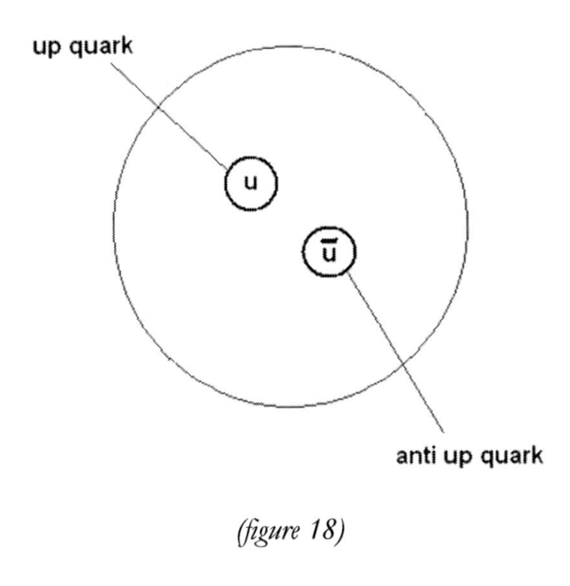

(figure 18)

This meson can also be referred to as 'uponium'.

Question:

One of the pi meson particles, of which there are three, consists of an up quark and an anti down quark. Use the quark fractional charge values to work out its charge now.

What did you get?

You should have reasoned out that it is a positive meson with a charge of +1, like the proton, and it is called the *positive pion*, or π^+ meson.

If you don't understand why, remember that the up quark has a fractional charge of +⅔, and a down quark has a charge of -⅓. However, since this is an *anti* down quark it must have the opposite charge, so it has a charge of +⅓ instead, thus giving the meson a charge of +1.

This meson is described below:

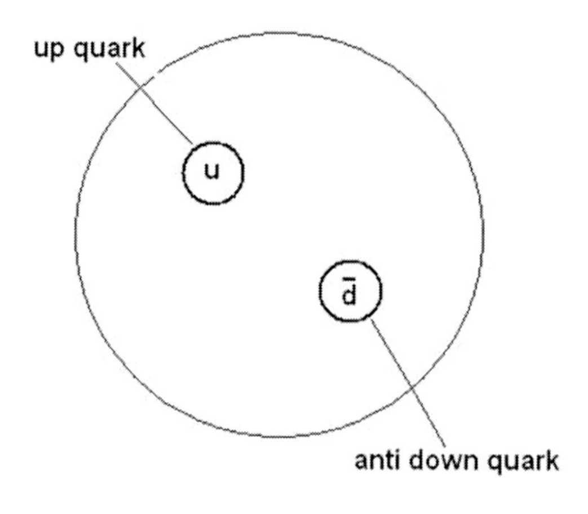

(figure 19)

There are three pi mesons and one of these is positive, one is negative, and one is neutral.

These are the pi plus (π^+), the pi minus (π^-), and the neutral pion (π^0) respectively. The lifetimes of both the negative and positive pions is the same at 2.6×10^{-8} second, or twenty six billionths of a second.

But the lifetime of the neutral pion is a mere 8.41×10^{-17} second which is about ten million billionths of a second.

This means that the neutral pion has a lifespan of only about a billionth of that of its charged siblings.

Pion Decay

The neutral pion as we know is unstable and it decays into a muon and a muon antineutrino according to the following reaction:

$$\pi^- \rightarrow \mu^- + \bar{\nu}_\mu$$

The muon then decays according to the equation we met in chapter 2.

As we already know, the muon then decays according to the equation into an electron, a muon neutrino, and an electron antineutrino.

But could the negative pion or the muon have decayed into any number of different particles?

Is there any way that we can predict what an unstable particle could end up decaying into, or at least what it could not decay into?

Maybe ……

Question

(a) By considering the mass of an up quark and a down quark, calculate the mass contribution from the quark masses to a proton's mass

(b) What is the mass of a proton?

(c) What percentage of a proton's mass is provided by the quark masses?

(d) What percentage of a proton's mass is 'missing'?

(e) Deduce from whence the deficit might come

CHAPTER 5: The Conservation Laws

1. Decay Processes

As we have learned, some particles are unstable, and these are in general particles which tend to have a *higher mass.*

Such particles usually have a very short lifetime, sometimes shorter than a millionth of a second, but exactly how do they decay?

Can we test if a decay that we suspect might happen is likely to happen?

Yes we can - in fact we can get a guide as to whether a particular suspected decay process might happen, and we can certainly rule out some decays that can't happen, and we do this by means of *conservation laws.*

These are laws which govern the natural decay of unstable particles, which can involve both hadrons and leptons, and if we suspect a decay process, we might want to test to see if that decay were possible.

These laws can give us a definitive answer as to which processes would be theoretically allowable and which would be forbidden.

Using these laws means that certain physical quantities must *balance* on each side of a decay equation.

An imbalance will completely rule out a suspected decay process, and a complete set of balances will give an indication as to whether or not a process looks likely, but isn't a cast iron guarantee.

1. Mass Energy Conservation

We learned in chapter three that mass and energy are equivalent, and any particle that does not possess relativistic mass can therefore be taken to be at rest.

Since, in that case, the total energy of the particle is bound up in its rest mass, it follows that the *total* energy of the products of any decay must add up to the total rest mass of the original unstable particle.

But remember that the rest mass of the original particle can manifest itself not only as the total rest masses of the product particles, but also as whatever kinetic energies those particles will possess, so that the rest mass of the original particle must find itself distributed as the rest mass *plus all other energies* of the product particles.

It further follows therefore that as long as the rest mass of the original unstable particle exceeds the total rest masses of the product particles, the decay is theoretically possible as far as mass energy conservation is concerned.

We need to know the rest masses of some of the array of particles, so here are the masses and lifetimes of some of our particles:

	Symbol	**Type**	**Rest Mass (MeV)**	**Lifetime (s)**
Proton	p	Baryon	938.26	Stable
Neutron	n	Baryon	939.55	880
Sigma minus	Σ^-	Baryon	1197.32	1.49×10^{-10}
Sigma plus	Σ^+	Baryon	1189.42	8.03×10^{-11}
Sigma zero	Σ^0	Baryon	1192.46	6.01×10^{-20}
Xi minus	Ξ^-	Baryon	1321.25	1.66×10^{-10}
Xi plus	Ξ^+	Baryon	2468.39	4.40×10^{-13}

Xi zero	Ξ^0	Baryon	1314.77	3.03×10^{-10}
Omega minus	Ω^-	Baryon	1672.55	8.2×10^{-11}
Omega zero	Ω^0	Baryon	2698.26	7×10^{-14}
Delta minus	Δ^-	Baryon	1232.17	6×10^{-24}
Delta plus	Δ^+	Baryon	1232.17	6×10^{-24}
Delta zero	Δ^0	Baryon	1232.17	6×10^{-24}
Lambda plus	Λ^+	Baryon	2286.46	2×10^{-13}
Lambda zero	Λ^0	Baryon	1115.61	2.6×10^{-10}
Negative kaon	K^-	Meson	493.67	1.24×10^{-8}
Positive kaon	K^+	Meson	493.67	1.24×10^{-8}
Neutral kaon	K^0_S	Meson	497.7	1.89×10^{-11}
Neutral kaon	K^0_L	Meson	497.7	5.2×10^{-8}
J/Psi	J/Ψ	Meson	3096.9	8.0×10^{-21}
Positive D	D^+	Meson	1869.4	10.6×10^{-13}
Neutral D	D^0	Meson	1864.6	4.2×10^{-13}
Negative pion	π^-	Meson	139.57	2.6×10^{-8}
Positive pion	π^+	Meson	139.57	2.6×10^{-8}
Neutral pion	π^0	Meson	135.0	8.3×10^{-17}
Neutral omega	ω^0	Meson	782	8.0×10^{-23}
Phi	Φ	Meson	1020	2.0×10^{-22}
Muon	μ^-	Lepton	105.7	2.20×10^{-6}
Electron	e^-	Lepton	0.511	stable

By way of example, let's consider the well known beta decay equation:

$$^{1}_{0}n \rightarrow \, ^{1}_{1}p + \, ^{0}_{-1}\beta + \bar{\nu}$$

We will now perform a mass balance:

Mass: (939.55) → (938.26) + (0.511)

The total mass after the decay = 938.77, and since this is less than the mass of the neutron (939.55) then the decay is permissible as far as mass energy is concerned. The equation also tells us that some 0.78 MeV of energy is distributed as kinetic energy amongst the products of the decay (excepting for the very small antineutrino mass).

2. Charge Conservation

Let's summarise first of all the charges on all the fundamental particles that we have encountered.

All leptons have a charge, Q, of -1, except for the lepton neutrinos which have a charge of zero, and all antileptons have the opposite charge to their lepton counterparts, this being +1, except of course for the antineutrinos which remain neutral.

As far as the quarks are concerned, the up, charm, and top quarks have a fractional charge of $+\frac{2}{3}$ and the down, strange, and bottom quarks have a charge of $-\frac{1}{3}$ and their antiquark doppelgangers have the opposite charge to them.

As far as the two stable baryons are concerned, the neutron is neutral and the proton has a charge of +1.

In any decay process or transmutation, the charge must be *conserved.*

This means the total charge before the process happens must be equal to the total charge after it has taken place.

Again considering the neutron decay equation:

$$^{1}_{0}n \rightarrow \, ^{1}_{1}p + \, ^{0}_{-1}\beta + \bar{\nu}$$

Applying charge conservation, and considering each particle in turn, what is the charge on all the particles involved?

The neutron is zero as is the antineutrino, and the proton has a charge of +1 with the electron having a charge of -1.

We now substitute the symbols for the particles by their charges, Q

$$^{1}_{0}n \rightarrow \, ^{1}_{1}p + \, ^{0}_{-1}\beta + \bar{\nu}$$

Charge: (0) → (+1) + (-1) + (0)

Hence the total charge on the left of the arrow, that is before the decay happens, is zero, and the total charge on the right of the arrow, after the decay has taken place, is also zero since the +1 and -1 charges cancel each other out.

It is clear therefore that charge has been conserved before and after the event, so this process is therefore permissible as far as charge conservation is concerned.

Besides the charge conservation law, there are others that we will now consider.

3. Lepton Number, L

Every lepton can be assigned a lepton number, L, and this number tells us if a particle is a either a lepton, an antilepton, or not a lepton at all.

All the leptons, including the lepton neutrinos, are assigned a lepton number of **L = +1**. Here they are:

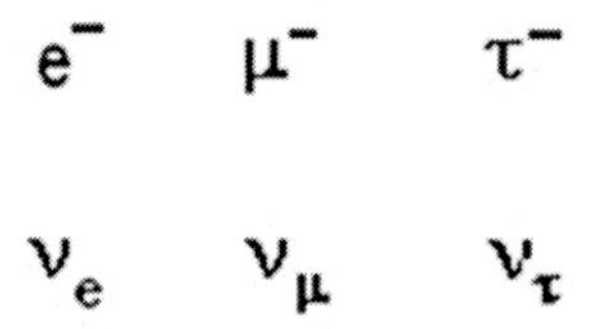

In addition to the leptons we have the antileptons, and these are each assigned the lepton number of **L = -1**.

Note that the antileptons contain the antilepton neutrinos as well.

Note also that in the case of the 'main' leptons, their anti lepton counterparts do not have to have a central line above their symbol to indicate that they are anti particles. Instead they need only to have their sign changed. For example, the anti particle for the electron, $\mathbf{e}^-$ is the positron, which is signified by the symbol $\mathbf{e}^+$.

Here are the antileptons:

$$e^+ \qquad \mu^+ \qquad \tau^+$$

$$\bar{\nu}_e \qquad \bar{\nu}_\mu \qquad \bar{\nu}_\tau$$

In the case of all non-leptons their assigned lepton number is **L = 0**.

4. Baryon Number, B

In a similar manner, every baryon can be assigned a baryon number, B.

This number will tell us if a particle is a baryon, an antibaryon, or not a baryon at all.

We will now assign baryon numbers to the quarks.

Surprising as it might seem, quarks do not have a baryon number of +1, as leptons did.

But all quarks are assigned a baryon number of **+⅓**.

This apparent anomaly is explained when we realise that a baryon itself, such as a proton or a neutron, will have a baryon number of **+1**, and this is because all baryons are composed of three quarks, the total baryon number of those three adding to **+1**.

Since quarks have a positive fractional baryon number, then the antiquarks will have a negative one, this being of course **-⅓**.

All non-baryons are assigned a baryon number of zero.

Let us now check that a lepton and baryon balance can be achieved with the beta decay equation:

$$^{1}_{0}n \rightarrow {}^{1}_{1}p + {}^{0}_{-1}\beta + \bar{\nu}$$

Lepton Number: (0) → (0) + (1) + (-1)

Hence lepton number is conserved

Baryon Number: (1) → (1) + (0) + (0)

Hence baryon number is conserved

Hence the four tests of mass, charge, lepton number and baryon number conservation have all balanced, so the indication is that this reaction is theoretically permissible, and we *know* that it is since it is a very well known reaction that has been observed many times.

Beta decay is a well known reaction which we know takes place, but what about decays that are forbidden by our four laws?

Let's now consider a hypothetical decay which suggests that a high energy proton could decay into a low energy proton together with a muon and a neutrino.

We will now apply our four laws to test this proposed decay:

We start with a mass energy balance:

$$ {}^{1}_{1}p \rightarrow {}^{1}_{1}p + \mu^{-} + \nu $$

Mass: (938.26) → (938.26) + (105.7) + (0)

Clearly there is insufficient mass on the left hand side of the equation to provide the masses on the right hand side, so this means that *the decay is forbidden.*

We will now use the charge conservation law to confirm that this decay is not possible:

$$ {}^{1}_{1}p \rightarrow {}^{1}_{1}p + \mu^{-} + \nu $$

Charge: (+1) → (+1) + (-1) + (0)

Since the total charge before the event is +1 and the total charge after the event is zero, since the negative and positive charges after the event cancel to zero, then charge is clearly not conserved.

This test has therefore confirmed to us that the decay suggested is forbidden and cannot occur.

2. Baryon Lepton Reactions

Thus far we have dealt with decay.

But besides the decay of unstable massive particles into lighter more stable ones, we have to deal with *reactions.*

Reactions occur when, for example, particles collide in a particle accelerator and as long as the energy of collision is sufficient, an array of new particles can be produced from the reaction.

We now have three conservation laws, these being conservation of charge, lepton number, and baryon number, and we can use these laws to determine if reactions are possible which can involve both leptons and baryons.

Note that we no longer need a mass balance as far as reactions are concerned.

Why is this?

The answer is that we now have another source of energy besides the rest mass of the particles that are about to collide, and that is the kinetic energies of those masses. Since energy and mass are equivalent, it does not follow that the rest masses of the products of the interaction must be less than the rest masses of the originating particles since the comparatively huge kinetic energies that can be reached in particle accelerators must be added to the total energy provided before the collision.

For example, let's test whether or not the following reaction is permissible in theory. We will, in turn, apply all three conservation laws to help us come to a conclusion. The reaction involves a muon neutrino which collides with a charm quark.

The masses are then converted into energy resulting in the proposed creation of an up quark, and a tauon.

We will apply lepton number conservation, baryon number conservation, and then charge conservation in turn:

$$\nu_\mu \quad + \quad c \quad \rightarrow \quad u \quad + \quad \tau^-$$

Lepton no., L:	(+1)	+	(0)	→	(0)	+	(+1)

This is so because the neutrino is a lepton with L = +1, as is the tauon, but the charm quark and the up quark are not leptons so have the number L = 0.

Since the lepton numbers on both sides of the arrow total +1, then lepton number is conserved.

Baryon no., B:	(0)	+	(+⅓)	→	(+⅓)	+	(0)

The two quarks are each assigned the baryon number +⅓ and the two leptons have baryon number zero. Since the baryon number on both sides of the equation is +⅓, then baryon number is also conserved.

So far the reaction seems permissible.

Charge, Q:	(0)	+	(+⅔)	→	(+⅔)	+	(-1)

The total charge before the reaction is +⅔, but it is -⅓ after it. This means that charge is not conserved, so this reaction is *not* permissible.

We now know how to test to see whether or not a reaction might be forbidden due to an imbalance of lepton number, baryon number, or charge, but can we go further?

Is it possible to actually identify an unknown particle that may appear from an event? In the next chapter we look at how this can be done …..

Questions

1. By considering the masses and lifetimes of the particles in the list at the beginning of the chapter, state which particles are matter antimatter pairs.

2. What are the baryon numbers of the following particles?

(a) A down quark

(b) an electron

(c) an anti charm quark

(d) a neutral lambda baryon, λ^0

(e) an anti tauon

(f) a bottom quark

(g) a proton

(h) an anti proton

(i) a positive pion, π^+

3. Use the four conservation laws to test if the following decays are theoretically possible or if they are forbidden. In the cases where they are forbidden, state the reason(s) why this is so:

(a) $\mu^- \rightarrow e^- + \bar{\nu}_e + \nu_\mu$

(b) $n \rightarrow p + \mu^- + \bar{\nu}_e$

(c) $\Sigma^- \rightarrow p + \pi^+$

(d) $\Omega^- \rightarrow \Xi^0 + \pi^-$

4. The following reaction involves a collision between a high energy muon neutrino and a bottom quark.

Use the conservation laws to determine whether or not the following reaction is theoretically possible:

$$\nu_\mu + b \rightarrow u + \mu^-$$

CHAPTER 6: Particle Identification

The aftermath of many interactions are recorded in particle accelerators and often baryonic particles are observed the identity of which isn't immediately clear.

However, sometimes it is possible to identify these.

This is done by use of the three conservation laws that we dealt with in the last chapter, but remember that a mass balance is not required in an interaction due to the additional kinetic energy in the collision.

However, we need another tool at our disposal as well.

This is the application of *hadronic quark structure* to the interaction.

By this I mean that not only must charge, lepton number, and baryon number be conserved before and after an event, but an analysis of the quark structure of all the constituent baryonic particles before the interaction must yield a balance with the quark composition of baryonic particles after it.

However, quark balance in this book will be limited to reactions which do not include weak force mediation which can result in a change of quark flavour.

We will limit ourselves to interactions involving baryons and mesons only.

In order to do this we first need a list of some of the many baryons and a list of some of the many mesons, together with the quark structure of each combination particle.

We start with a list of baryons:

Name	Symbol	Quark structure
proton	p	u u d
neutron	n	u d d
sigma minus	Σ^-	d d s
sigma plus	Σ^+	u u s
sigma zero	Σ^0	u d s
Xi minus	Ξ^-	d s s
Xi plus	Ξ^+	u s c
Xi zero	Ξ^0	u s s

Name	Symbol	Quark structure
omega minus	Ω^-	s s s
omega plus	Ω^+	s c c
omega zero	Ω^0	s s c
delta minus	Δ^-	d d d
delta plus	Δ^+	u u d
delta zero	Δ^0	u d d
lambda plus	Λ^+	u d c
lambda zero	Λ^0	u d s

Now before you read any further, look carefully at this list and see if you can see anything that doesn't seem quite right, and in addition, if you spot anything, write down all the apparent anomalies that you can find.

What did you find?

What seems strange at first glance is that there are some baryons in the list that have the same quark structure.

How many did you find?

There are six baryons in three sets that show this, and these are the proton (p) and the delta plus (Δ^+) which both have a quark structure of (u u d).

Then the neutron (n) and the delta zero (Δ^0) both have the same quark structure of (u d d).

Finally the sigma zero (Σ^0) and the lambda zero (Λ^0) both have a quark structure of (u d s).

How can different baryons have the same quark structure?

The answer is that one of each pair has a greater energy than the other one and is therefore a more massive 'version' of its lower energy twin, and this will very quickly decay into its more stable twin.

However do not automatically assume that the lighter particle is stable.

The lighter one may be unstable too!

Of the three pairs that we have isolated, the more energetic particles are the lambda plus (Δ^+), the delta zero (Δ^0) and the sigma zero (Σ^0) respectively.

Let's remind ourselves of their rest masses and their lifetimes:

Particle	Rest mass (MeV)	Lifetime (s)
Proton, p	938.26	Stable
Delta plus, Δ^+	1232.17	6×10^{-24}
Neutron, n	939.55	880
Delta zero, Δ^0	1232.17	6×10^{-24}
Sigma zero, Σ^0	1192.46	6.01×10^{-20}
Lambda zero, Λ^0	1115.61	2.6×10^{-10}

Notice two things about these baryons – firstly the lambda plus, the delta zero and the sigma zero are more massive than the proton, neutron and lambda zero into which they respectively decay.

Secondly the more massive baryons have shorter half lives than their lighter counterparts.

Besides the baryons, we also now need a list of some mesons:

Name	Symbol	Quark structure
K minus	K^-	$s\bar{u}$
K plus	K^+	$u\bar{s}$
K zero	K^0	$d\bar{s}$
J/Psi	J/Ψ	$c\bar{c}$
D plus	D^+	$c\bar{d}$
D zero	D^0	$c\bar{u}$
Pi minus	π^-	$d\bar{u}$
Pi plus	π^+	$u\bar{d}$
Pi zero	π^0	$u\bar{u}$ or $d\bar{d}$

Name	Symbol	Quark structure
Omega zero	ω^0	$u\bar{u}$ or $d\bar{d}$
Phi	Φ	$s\bar{s}$

Let's now look at an example of how to identify an unknown particle using the conservation laws and an analysis of the entire quark structure.

Examples

1. An event in a particle accelerator reveals that a lambda zero particle colliding with an unknown particle, X, produces a proton.

Identify the unknown particle.

Solution

The interaction equation is given as:

$$\Lambda^0 + X \rightarrow p$$

Charge, Q: (0) + (X) → (+1)

Hence, X must have a charge of +1

Lepton number, L: (0) + (X) → (0)

Hence X cannot be a lepton

Baryon number, B: (1) + (X) → (1)

Hence, since the baryon number is already balanced, X must be a positive meson.

We now perform a quark structure analysis by referring to our tables above:

Quark structure: (u d s) + (X) → (u u d) *(equation 1)*

Now the quark composition on both sides of the arrow, that is both before and after the reaction, must balance.

There are two ways to do this, one of them being to check that a particular quark before the reaction is seen after the reaction on the right hand side of the arrow.

The other is to check if a quark on any side of the arrow can be eliminated by an antiquark of the same flavour on the *same* side of the arrow.

In *equation 1* above, the up quark and the down quark before the reaction are accounted for after the reaction so they are eliminated from the analysis.

But notice that there is an *extra* up quark present after the reaction so that up quark must appear before the reaction in our quark box for particle X.

In addition to that, there is an 'extra' strange quark present before the reaction and the only way to make that balance in the equation is to insert an *anti* strange quark in the quark box for X before the reaction, giving us the following solution:

Quark structure: $(u\ d\ s) + (u\ \bar{s}) \rightarrow (u\ u\ d)$

Now, checking our table of mesons, what is X?

Our list tells us that X must be a K^+ meson.

2. (Try the following example on your own before looking at the solution). In a particle accelerator, a proton is collided at high energy with a K^- particle. Two resulting particles are observed after the event, one of them, the particle Z, being unknown. Identify the unknown particle.

The reaction is given as:

$$\mathbf{K^- + p \rightarrow Z + \pi^+}$$

Charge, Q: $(-1) + (+1) \rightarrow (Z) + (+1)$

Hence for charge to be conserved, Z must have a charge of -1

Lepton number, L: $(0) + (0) \rightarrow (Z) + (0)$

Hence Z cannot be a lepton

Baryon number, B: (0) + (1) → (Z) + (0)

(since the π^+ and K^- are mesons and the proton is a baryon)

Hence Z must have a baryon number of +1 in order to balance the equation, meaning that Z must be a negative baryon.

Quark structure: $(s\ \bar{u})$ + $(u\ u\ d)$ → (Z) + $(u\ \bar{d})$

One up quark from the proton balances with the up quark in the π^+ meson and the anti up quark in the K^- meson cancels with the other up quark in the proton.

So in the brackets for Z we must insert a down quark to balance with the down quark seen in the proton and another down quark to cancel the anti down quark in the π^+ meson.

We also need one strange quark to balance with the strange quark in the K^- meson.

This gives the quark structure of the unknown negative baryon, Z, as (d d s).

Hence by looking at our table of baryons we seen that Z must be a $\boldsymbol{\Sigma}^-$ baryon.

Questions

Identify the unknown particle **'X'** in each of the following theoretical reactions:

1. $n + \Sigma^0 \rightarrow p + (X)$
2. $\Sigma^0 + \pi^- \rightarrow \Xi^- + (X)$
3. $(X) + p \rightarrow n + \pi^0$
4. $\Delta^0 + (X) \rightarrow \Lambda^+$

CHAPTER 7: Particle Accelerators

In relatively recent decades many new and exotic particles have been discovered, some of which were predicted to exist and some of which weren't.

So what is a particle accelerator?

Perhaps the simplest is the cathode ray tube found in older T.V. sets and oscilloscopes.

This is an evacuated tube, along which particles are accelerated with the aim of colliding them with the screen at the end of the tube, the part of the tube that we see. This is coated with a phosphorescent material such as zinc sulphide which will give off visible light when a collision takes place.

The particles accelerated in these are electrons:

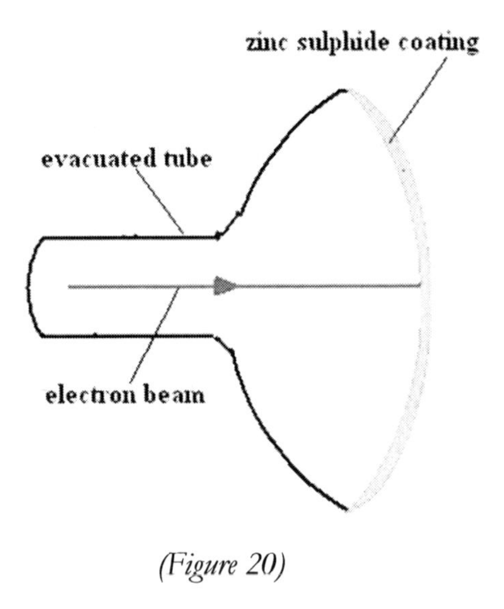

(Figure 20)

The electrons can reach a speed in the order of a million metres per second when they hit the zinc sulphide.

What sort of energy level can the electrons in such a tube gain?

When they strike the screen they have achieved an energy of about 30,000 eV, which is 0.03 MeV, or 0.00003 GeV.

Bear this energy level in mind as we progress through the energy levels in some of the big experimental accelerators.

But particle accelerators which are built to find new particles are grander affairs than this.

The energy of the particles that are accelerated in experimental machines is such that a total annihilation will occur resulting in pure energy out of which new particles will be produced.

We will categorise these experimental accelerators as either:

(a) Linear Accelerators (Linacs) or,

(b) Circular Accelerators (Cyclotrons and Synchrotrons)

The Linear Accelerator (Linac)

Linear accelerators or Linacs accelerate charged particles by synchronising the timing of an alternating voltage such that any particle is always accelerating, regardless of its position in the accelerator tube.

Particles will be accelerated through a series of *drift tubes* which are held at a reversible potential such that as the particle approaches the tube, the potential at the end of the tube it has just left is the same as that of the particle.

This means that repulsion occurs, while the potential at the end of the drift tube that it is approaching is the opposite polarity so that attraction occurs there.

When the particle passes through each tube its potential is reversed so that the beam of charged particles is always accelerating:

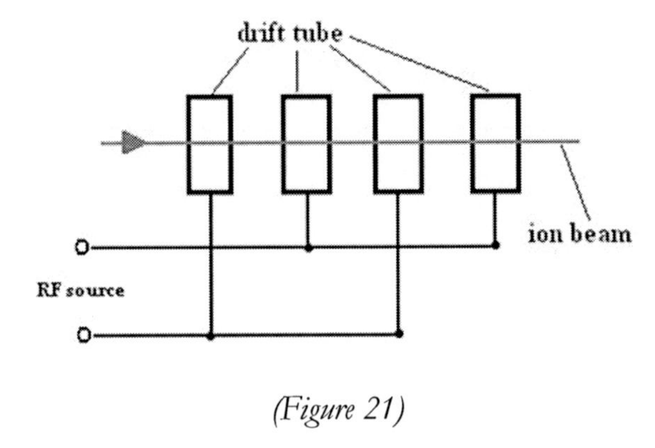

(Figure 21)

Clearly the longer that a linac is, the greater is the speed to which the particles that make up the ion beam can be accelerated, so when they arrive at the target they have a higher energy.

As time has passed, particle physicists have built machines with ever greater energies available at the collision point.

Why is this?

The reason is that greater energies are needed to try to create the more exotic particles with higher masses.

A typical collision energy for a linac might be 50 MeV.

Can you now work out how many times greater would be the energy of a particle at the collision point in such a machine compared with the humble T.V. tube?

It comes out at around 1667 times that energy!

The largest linac in the world at the present time is at Stanford University in the U.S.A.

This is known as the Stanford Linear Accelerator Centre (SLAC) and its linac is two miles or 3.2 km long! This is what it looks like from the air and notice that a major freeway crosses the linac:

(Photo courtesy of SLAC National Accelerator Laboratory)
(figure 22)

Construction started in 1962 and was completed in 1966 when research started.

What sort of energies can be achieved at SLAC?

In 1989 it became possible to accelerate electron and positron beams which were crashed into each other at energies of 50 GeV.

By way of comparison, can you work out how much greater this is than the energy of the electron beam in a T.V. tube?

It comes out at about 1.67×10^6, or 1.67 million times greater!

This linac is famous for a number of reasons, two of which are:

(a) deep inelastic scattering

(b) the discovery of new particles

Deep Inelastic Scattering

It was in 1961 that Murray Gell-Mann first postulated the idea that protons and neutrons were not fundamental particles and that they were composite particles composed from quarks.

It took another seven years for this idea to be vindicated and in 1968 the first evidence for the existence of quarks was seen.

What sort of experiment was necessary to test for the existence of quarks?

Remember that in order to investigate what was in an atomic nucleus Ernest Rutherford had to fire alpha particles at gold nuclei.

What was pertinent about that?

The answer is that alpha particles were small enough to actually penetrate the comparatively large gold atom.

So it follows that in order to investigate the interior of a proton, something much smaller was needed to penetrate protons.

Can you think of a suitable particle to do the job?

The answer is of course the electron which is only 5.4×10^{-4} times the proton mass.

If the quark model proved to be wrong, then an electron which made a close pass to or even penetrated a proton might be expected to be attracted to its centre of charge, like this:

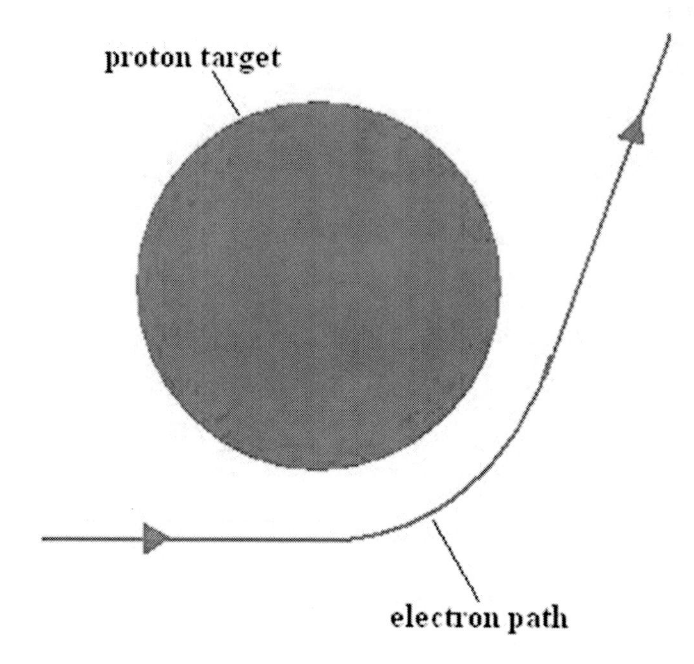

(Figure 23)

However, analysis revealed that the electrons were observed to be deflected away from the centre of charge, like this:

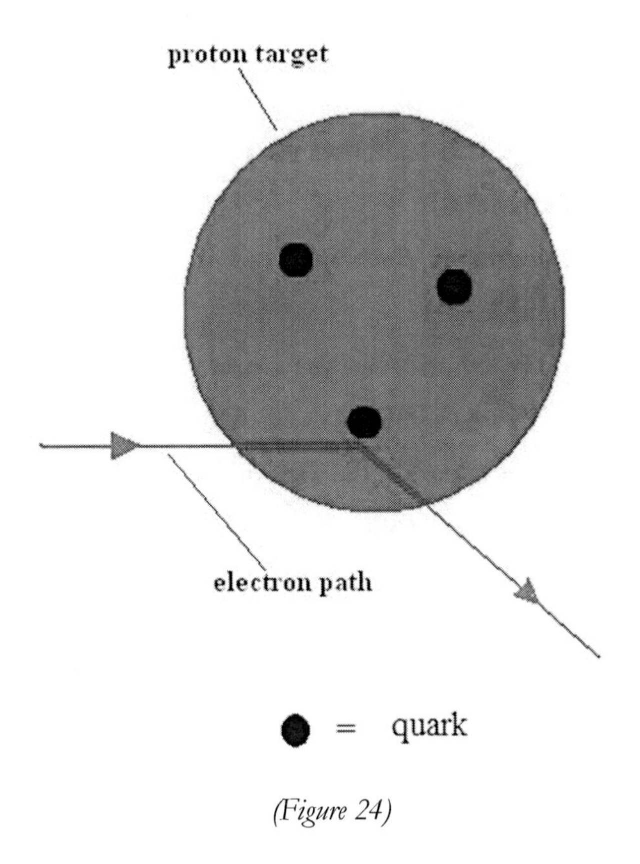

(Figure 24)

This could only mean one thing – that there was something with mass within the confines of the proton that the electron had impacted with and which had caused the deflection.

This process is called *deep inelastic scattering* and it was done in the SLAC because the very high electron energies to enable proton penetration was available.

Discoveries at SLAC

In the 1970s, work was being undertaken in SLAC with the Stanford Positron Electron Asymmetric Ring (SPEAR), in which fast moving electrons were collided into their antiparticle, the positron.

This of course resulted in annihilation, with an accompanying release of energy, and new particles were created.

In 1974 the J/Ψ meson was discovered and in 1976 the charm quark and the tau lepton were confirmed.

D mesons were observed, and this was a confirmation of the discovery of the charm quark, since D mesons contain a charm quark.

Physicists working on some of the experiments in SLAC were awarded the Nobel Prize in physics for their work.

The particles which were accelerated reached speeds in excess of 99% light speed in the accelerator, but how could higher energies be achieved?

Clearly the answer would have been to build ever longer accelerators, but there is a limit to this.

One way to cut down on the length of linacs is to build *circular* accelerators instead, thereby doing away with a linear track altogether.

Since the application of a magnetic field perpendicular to an ion beam will cause the beam to describe a circular arc, all that is needed is a means of accelerating the ions together with the application of such a field.

The Cyclotron

Cyclotrons consist of two 'D' shaped chambers which are appropriately called 'dees'. A magnetic field is applied to the dees to cause circular motion of the ion beam, but note that the ions do *not* undergo any acceleration whatsoever *within* the dees.

In other words, although they are accelerating in that they are changing their direction (because acceleration is a vector quantity), they are not gaining speed there.

But they clearly need to increase their speed, and this takes place in the short gap *outside of* and *between* the dees in which a constant frequency alternating voltage is applied each time they make a pass, and such that the polarity is always reversed in phase with their passage and is always such that attraction to the next dee occurs.

This occurs with the radius of the arc within the dees increasing with each speed gain until the ions eventually exit the machine.

The principle of the cyclotron is thus:

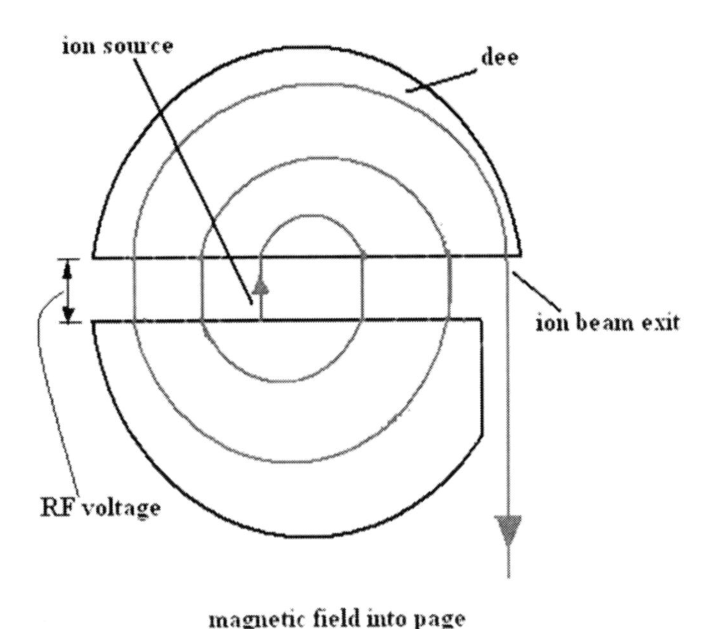

(Figure 25)

Obviously the length of a linac can be squeezed into a small area with such a machine and this is the advantage, but are there any disadvantages?

Yes – notice that the alternating source is constant frequency.

But as the ions accelerate to relativistic speeds, their arrival at the dees tends to drift out of phase with the alternating source, and this usually puts a limit on the speed achievable by the ions of around 5% 'c'.

The Synchrotron

Synchrotrons are circular accelerators as well, but they avoid the phase drift problems generally found with the cyclotron.

Can you guess how this is achieved?

It is done by varying both the alternating voltage and the magnetic field strength along the track so that the ion beam can achieve greater speeds and hence greater energies.

Synchrotrons are much larger than cyclotrons, and we will be talking about the world's largest one later.

Things have moved on almost at breakneck speed in the world of particle accelerators since the 1980s.

Then typical characteristics would have been currents in the order of milliamps with energies to a maximum of gigaelectronvolts and with particle densities of about a million billion particles per cubic centimetre whereas now they can reach energies of teraelectronvolts (1×10^{12} eV), integer amp currents and particle densities ten million times greater!

Question

Look at this equation in which a high energy photon collides with a proton in a particle accelerator experiment:

$$\gamma + p \rightarrow x + \bar{D}^0 + n$$

In the reaction an anti D meson and a neutron are observed, plus an unknown particle. Use the conservation laws and quark structure analysis to identify **X**

CHAPTER 8: Detection

How can particles that are created in a particle accelerator be detected?

Detection is dependent on *ionisation.*

Fast moving charged massive particles will readily ionise the atoms in the medium through which they are passing immediately after their creation.

Once the medium is ionised, the particles can be tracked by either *potential change* or *gas induction.*

1. Potential Change

Just as we can't necessarily see a high flying jet, we can almost always track it by its vapour trail, so although we can never 'see' the actual particle we can track its movements and what it does by the ionisation of the medium of transit.

So what does it mean to ionise the medium of transit?

Do you remember how a charged electroscope can be discharged by the presence of a radioactive source?

In the following diagram, the electroscope is negatively charged and the alpha particles are positive:

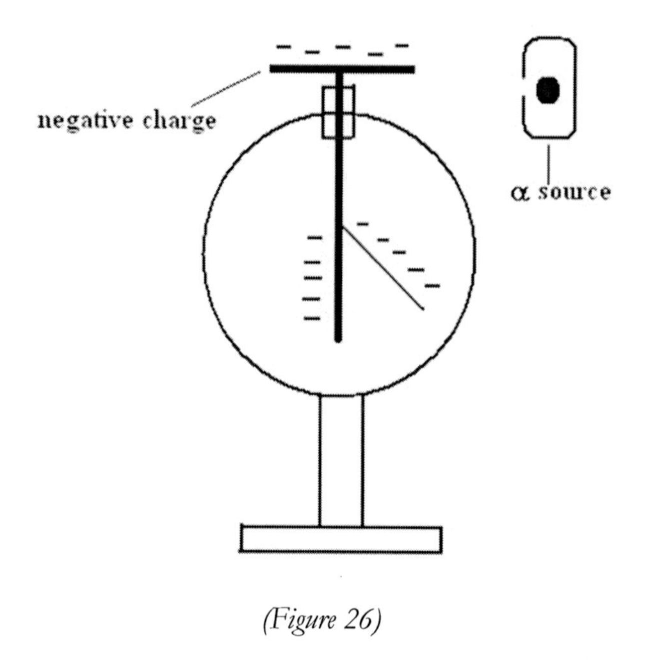

(Figure 26)

Don't get confused as to why the electroscope discharges – it isn't the positive alpha radiation that neutralises the negative charge on the top plate, but rather it is that the alpha radiation, which is one of the most ionising of radiations, will ionise the air and it is the positive ions in the air surrounding the electroscope that are attracted to the top plate and which then do the neutralising, with the result that the gold leaf will fall.

Ionisation of a gas is the principle on which the Geiger-Muller tube works.

The tube is filled with an inert gas like argon and a central electrode is held at a potential of about 400 volts with respect to the tube sides:

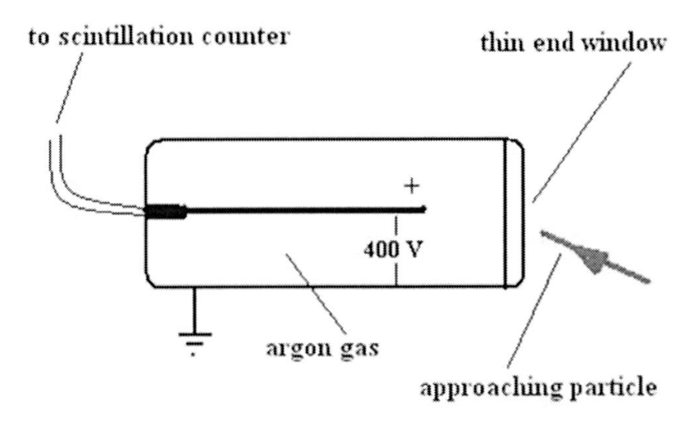

(Figure 27)

When an ionising particle enters the tube via the thin end mica window, the argon gas becomes ionised and ionised electrons will be attracted to the central positive electrode.

As they move towards it they will, in turn, ionise other atoms and more free electrons will be present and they will be in sufficient quantity to cause a momentary drop in voltage, which is recorded on the scintillation counter.

The release of more free electrons in transit is called the *avalanche effect.*

The Drift Chamber

To detect particles after events in accelerators, drift chambers can be used which depend for accurate operation on the avalanche effect, but don't get these confused with the drift tubes in linacs.

Whereas the Geiger-Muller tube has one central electrode, a drift chamber will have a series of electrodes which can be placed a few centimetres apart, which are held at a positive potential of a few thousand volts above the grid plates through which an incoming particle can pass.

This potential is calculated to be below the level at which a spontaneous electronic discharge can occur between electrodes.

As a result the gas between the grid and the electrodes will be ionised and negative ions can cause an avalanche effect and will be attracted to the nearest positive electrode.

This enables a picture to be built up of where particles have been, how fast they were travelling and how long they were around for:

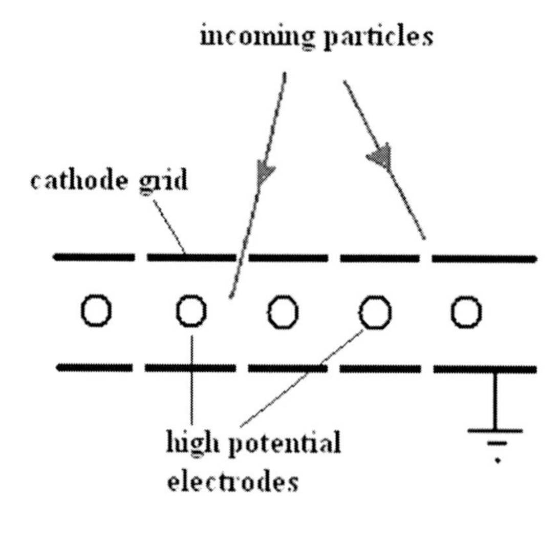

(Figure 28)

Silicon Detectors

Silicon detectors use semiconductor materials to record the passage of particles and they can provide a more accurate picture of the paths of particles that penetrate them. Typically a p-type/n-type diode junction is used which is set in *reverse bias*, so that no current passes.

But any particle which then passes through the junction will cause an ionisation current which is easily detectable:

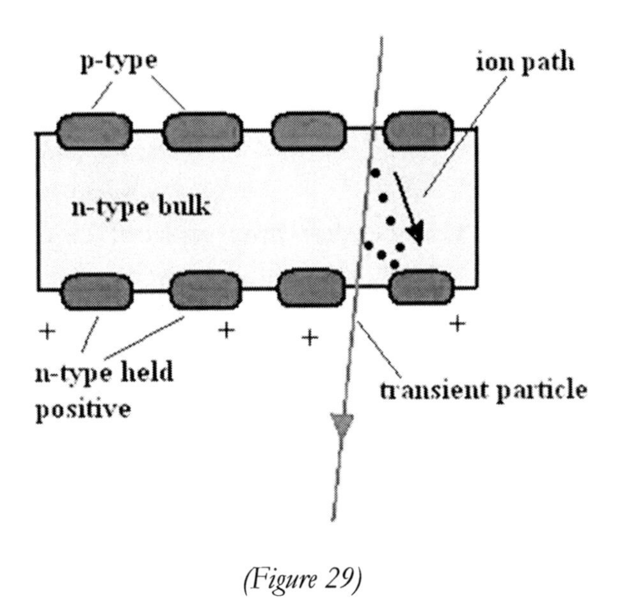

(*Figure 29*)

If an arrangement of thousands of such junctions is set up around a collision point, then the path and history of a transient particle can be very accurately recorded. However, silicon detectors are considerably more expensive than drift chambers.

The high cost is due in part to the cooling system which is necessary to reduce noise to acceptable levels.

Silicon detectors are used in experiments in the Large Hadron Collider in CERN in Geneva.

2. Gas Induction

Gas induction is generally used to monitor particle creation and decay in fixed targets. What do you think would be a suitable fixed target particle which is readily available in which energetic particles could be rammed into?

The answer would be protons which are readily available in hydrogen, since hydrogen nuclei are protons, so that any high energy particle entering hydrogen would be bound to encounter a proton.

The Bubble Chamber

By and large the bubble chamber has been replaced in the large and more sophisticated accelerators of today by silicon detectors, but the principle is important especially for describing principles of particle analysis and identification.

Bubble chambers can hold liquid hydrogen, usually held at the boiling point of hydrogen which is -253°C or 20K.

When a liquid boils bubbles obviously appear, but could there be a way to make bubbles appear *only* when a particle transits through the hydrogen?

Yes there is, but first call to mind the fact that the boiling point of liquids reduces at low pressure, so that to boil an egg on the summit of Mount Everest takes a lot longer than at sea level.

This is because the boiling point of water is reduced to a mere 69°C with the result that it takes longer for heat transfer from the water to the egg.

But the same principle is applied in reverse, in that liquids under increased pressure have a raised boiling point, with the result that the old 'pressure cookers' would do their job faster since the temperature at which water boiled was increased and hence heat transfer was speeded up.

Now the liquid hydrogen in a bubble chamber is normally kept at a pressure of about five atmospheres so that no boiling would normally occur because its boiling point is now raised.

However, if the pressure were reduced back to one atmosphere, then boiling could readily occur, but one other ingredient is necessary for that to happen.

That is that a bubble needs a nucleation point to form, and without such no bubbles will form.

What better nucleation point could there be than the ions created in the hydrogen by the passage of a transient ionising particle!

So if the pressure could be reduced so that normal boiling would occur at its ambient temperature, and ionisation took place, then bubbles will form.

The pressure can be reduced by holding the liquid nitrogen above a moveable piston, like this:

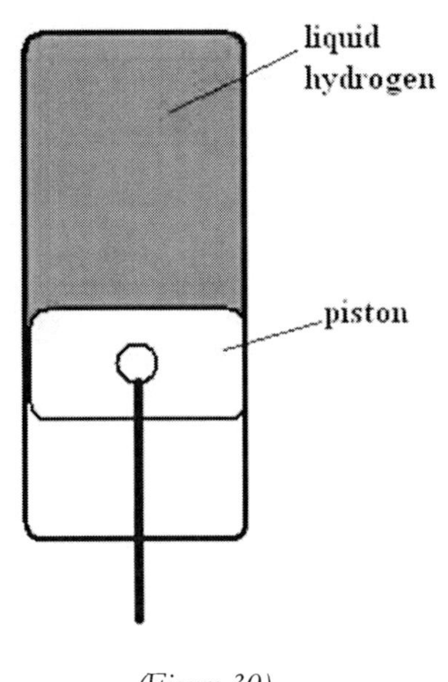

(Figure 30)

This is exactly what happens in a bubble chamber, and pressure reduction by the movement of a piston is made to happen simultaneously with a planned particle event, so that the path of particles and other information about the event can be seen by the bubble tracks.

Both the incoming particles and those created after the event can be monitored in this way and photographs can be taken of the bubble tracks.

There is one drawback however, and that is that *neutral* particles will not ionise and hence will leave no bubble trace.

This is due in large part to their lack of charge which would influence electrons on atoms within the liquid to become ionised.

There are a number of ways in which we can get some clues as to what particle a given track might represent.

One of these is to look for heavier, denser tracks, which if seen would indicate that the particle was heavily ionising and slow moving, thus allowing it to ionise more target atoms in its path.

How can we differentiate between oppositely charged particles in the chamber?

This can be achieved by making charged particles curve and this in turn is done by the application of a magnetic field perpendicular the direction of motion of the particle, although there is no guarantee of perpendicularity to any particular particle that is being observed.

As a result any charged particle that is not quite travelling perpendicularly to the field will describe a helix:

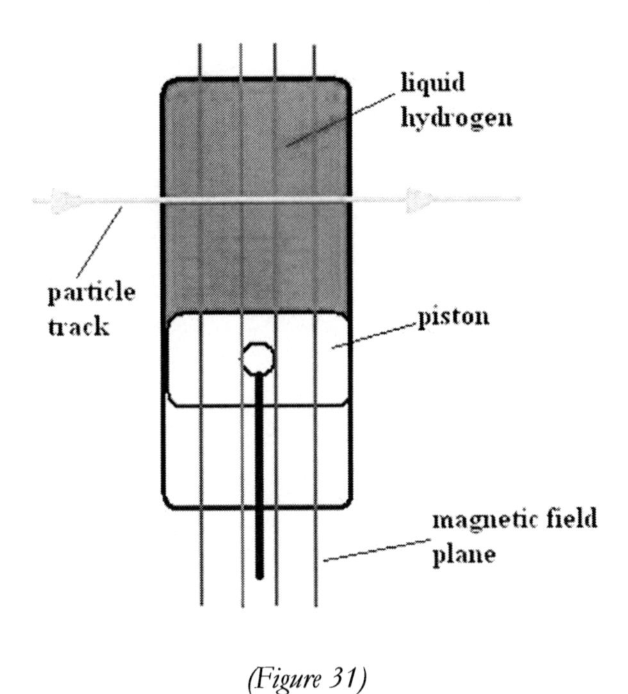

(Figure 31)

In the early 1970s the Big European Bubble Chamber (BEBC) was installed at CERN, Geneva, and this vessel contained 35 cubic metres of liquid hydrogen and deuterium.

The superconducting coils of the magnet for the 3.7m Big European Bubble Chamber needed to be checked.

Look at the size below:

(Reproduced by kind permission of CERN – CERN-PHOTO-7407001 ©CERN Geneva)
(figure 32)

By December 1974, the magnet had reached the field design value of 3.5 Tesla (T).

Compare this magnetic field strength now with that of the Earth's flux, which is in the order of 0.00005 Tesla (T).

One of the massive pistons is shown here:

(Reproduced by kind permission of CERN – CERN-PHOTO-7608011 ©CERN Geneva)
(figure 33)

By the end of its life in 1984 the BEBC had produced some 6.3 million photographs of particle interactions.

The whole process of taking a photograph of a particle interaction is carried out in a very short space of time – in the order of a few milliseconds.

Question

Gamma radiation, which is neutral, has the ability to slightly ionize when passing through a cloud chamber.

Explain why a neutral particle is less likely to cause ionization in a bubble chamber and why a charged particle will readily ionize

CHAPTER 9: Event Analysis

Photographs of particle interactions can be very complex, as you can see from the following photograph of 16 GeV π^- mesons entering the picture from the left in a bubble chamber:

The decay from the 16GeV pions was captured in CERN's first liquid hydrogen bubble chamber and the decay products can be seen spiralling due to the magnetic field perpendicular to the picture.

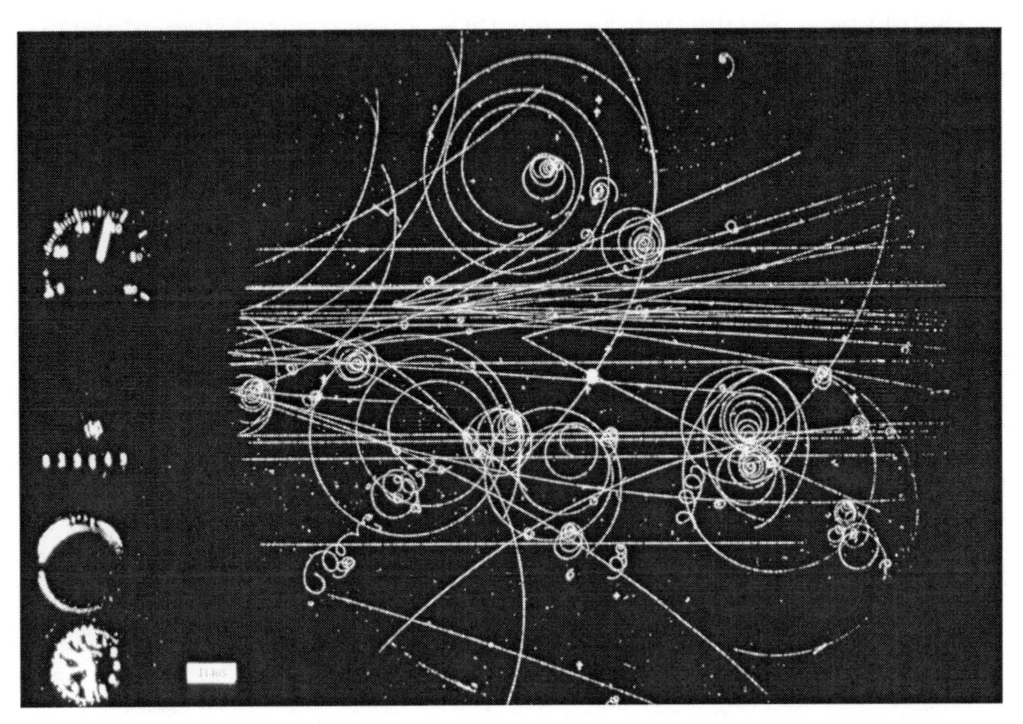

(Reproduced by kind permission of CERN – CERN-EX-11465 ©CERN Geneva)
(figure 34)

It is of course possible for two particles to collide with each other and simply bounce off each other without annihilation into energy. In this case no new particles will be produced. Can you think of a way that annihilation would be guaranteed with the subsequent creation of new particles?

The answer is to collide a particle and its antiparticle, such that annihilation will always happen on contact, and a subsequent release of energy from which new particles may be formed is bound to happen.

So a very widely used process in particle accelerators has been to collide a particle with its antiparticle doppelganger.

This can be an electron with its antiparticle, a positron, and amongst the new particles created it is common to observe new particle/antiparticle pairs created.

Before the installation of the new Large Hadron Collider in CERN in Geneva, the tunnel was used to smash a beam of matter particles into a beam of their antiparticles. This was the LEP collider, which stands for Large Electron Positron collider.

But it is also possible to collide other matter/antimatter pairs into each other.

For example, protons and antiprotons can be used.

A typical proton/antiproton collision might yield an array of new particles created such as that shown below:

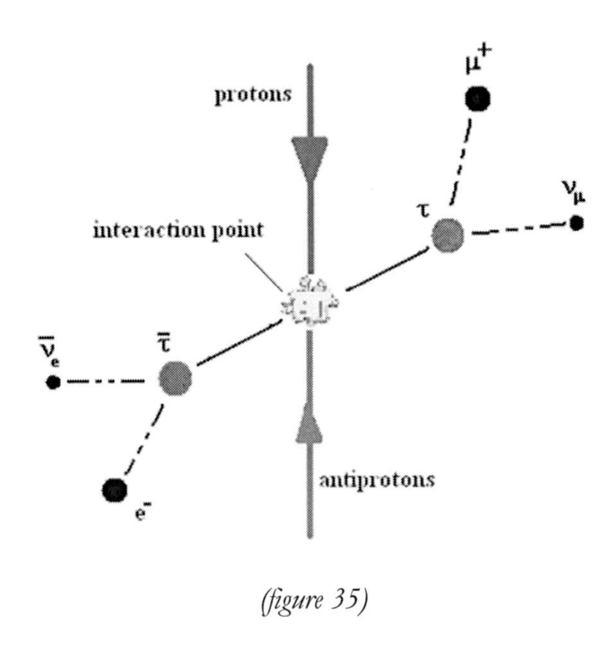

(figure 35)

Matter Antimatter Pair Recognition

How can a matter antimatter pair of particles created after the interaction be recognized?

As long as a magnetic field was present, a pair of tracks, one for the matter particle and one for the antimatter particle, would be seen, as long as the pair were charged, but they would both describe tracks *in opposite directions.*

Furthermore the tracks would be circular, and they would have the same radius, but only as long as they each had the same energy.

Questions

1. Look at the following diagram in which a magnetic field is flowing perpendicular to the page and which shows two circular tracks migrating away from the interaction point just after their creation.

State whether or not you think that the two circular tracks could be a matter antimatter pair and don't look at the solution until you've made a judgement:

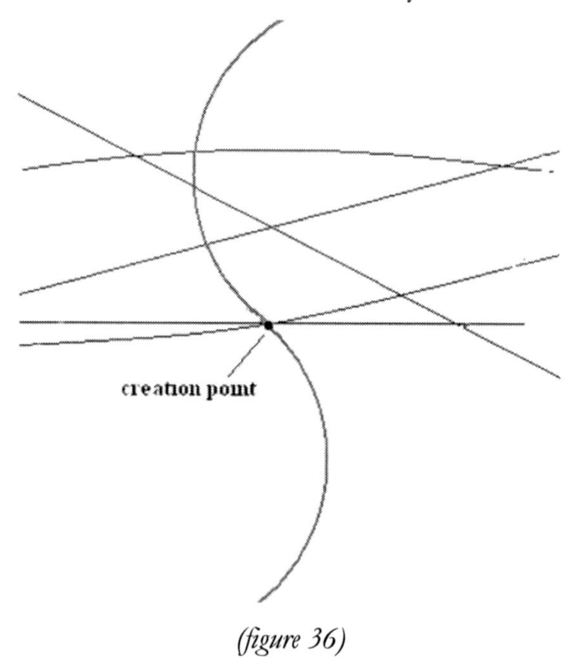

(figure 36)

What did you decide?

Check your answer in the answers and solutions section.

2. Look at the diagram below which shows part of an interaction, and say what you can deduce about the events that take place at point 'X'

What can you say about the two outermost tracks after the event at 'X'?:

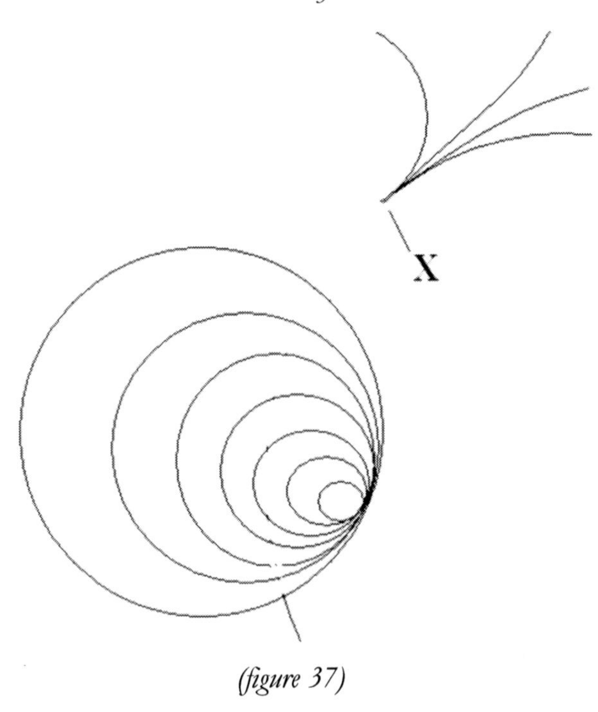

(figure 37)

Check your answers in the answers and solutions section.

3. The following is a representation of a bubble chamber photograph in which a negative kaon (K^-) collides with a stationary target proton with sufficient energy to cause annihilation and production of new particles including a low energy proton and a negative pion (π^-):

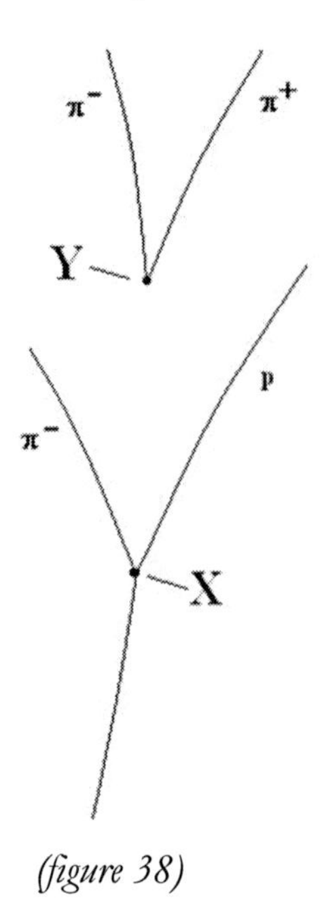

(figure 38)

(a) State what else happens at point X

(b) What happens between points X and Y?

(c) What happens at point Y?

(d) What can you say about the particles produced at point Y?

(e) Write the equation describing the event and identify the unknown particle produced at point X

Check your answers in the answers and solutions section

CHAPTER 10: Nature's Forces

Maybe you think that there are many forces in the world around you like the force from the wind, the force from a jet engine, tension forces, compression forces, thermodynamic forces, lift, drag, and the list can be very long.

We can think of these in general terms if we wish, but that is not what we mean by nature's forces.

Actually there are only *four* fundamental forces in nature and we are limiting our discussion to these four forces.

The reason that we have now to look at these four forces is that there is still an array of particles called *gauge bosons* that we haven't met yet, and they are responsible for operating in harmony with each of nature's four forces.

As we know, like charges repel and unlike charges attract.

Now, remember the composition of a proton – it contains two up quarks and one down quark, the down quark having a fractional negative colour charge of -⅓ and the up quark having a fractional positive colour charge of +⅔.

The down quark will be attracted to an up quark since they are oppositely charged, but the two up quarks will be repelled from each other by a greater amount due to their greater like charges.

So how do we answer this question; "why don't the two up quarks separate and fly apart?" – after all there seems to be nothing tangible that sticks them together – no duct tape, no pins and no micro-nuts and bolts.

If the boundary circle that we use in this diagram has no practical meaning at all:

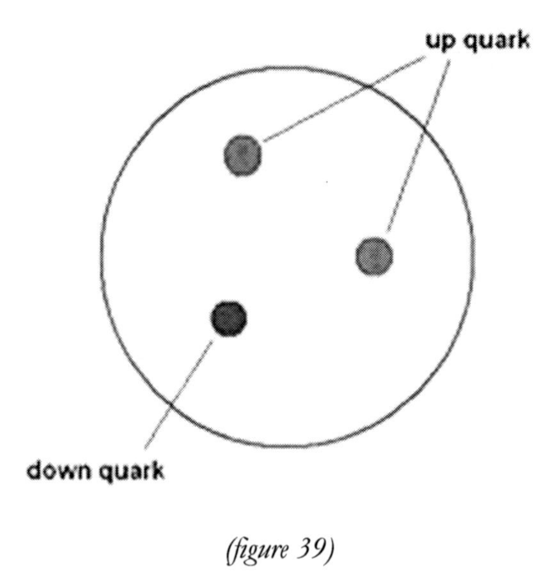

(figure 39)

Then in reality the following diagram might be more appropriate:

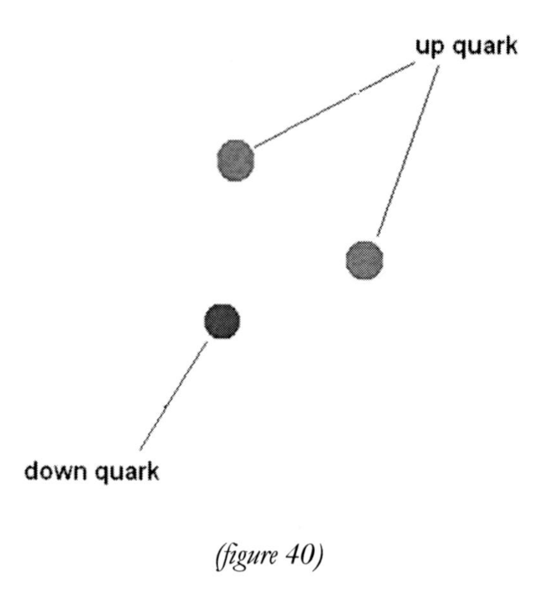

(figure 40)

So why are the up quarks staying together?

And what is more, why are they staying so close to each other, because they are separated to a distance of less than 1×10^{-15} metre which is the diameter of a small nucleus?

Clearly there has to be something stopping them flying apart and there is!

Actually the boundary in the diagram *does* have some meaning and it happens to represent the limits of action of nature's strongest force with the appropriate name of:

1. The Strong Nuclear Force

The strong nuclear force is responsible for overcoming the huge repulsive forces that are present due to the colour charges of identical quarks.

These repulsive forces are comparatively great at the very small distances at which quarks operate together, so it needs a very strong counter force to overcome them. The strong force acts only on quarks so that leptons cannot feel is presence at all.

There are some surprising things about the strong nuclear force.

The first one is that it has a very small range of operation, this being the diameter of a small nucleus of 1×10^{-15} m, and outside of this range it has no influence.

But even so the range of the strong force is some one thousand times greater than the notional diameter of an up or down quark, which is taken to be in the order of 1×10^{-18} metre.

The second property of the strong nuclear force which is *unique to that force* is that it acts something like a rubber band!

If you have stretched a rubber band with your hands you will have noticed that the further you stretch it, the more effort it takes.

This is not surprising because, although it may not obey Hooke's law perfectly due to its long chain molecules unravelling in a less than perfect way, it does approximate to the way a spring behaves.

But the problem is that when it is at the end of its stretching capability, when its long chain molecules have unravelled, it takes an even greater force to stretch it further and when it does it breaks!

The strong force behaves in a similar, but not identical, manner.

Its efficacy increases with distance – in other words, if two up quarks in a nucleus were to begin to separate, then the strong force would act with greater strength to bring them back together again, something like this spring analogy:

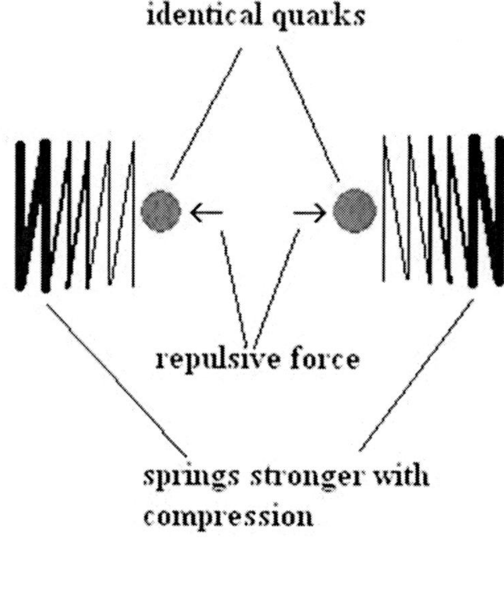

(figure 41)

Of course, if the quarks were to separate to outside of the range of the strong force, then it would no longer have any effect on them.

Actually quarks in a hadron appear to be able to move about quite freely when they are in close confinement, and it is only when they begin to separate that the energy needed to overcome the strong force increases at a rate of about 1GeV per femtometre (1×10^{-15} m).

Since the effect of the strong force asymptotically tends to zero as the quarks get closer, the process of how the strong force behaves in a hadron is referred to as *asymptotic freedom.*

But remember that quarks do not congregate individually and they are always found bound together in hadronic particles.

But there is another example of charge repulsion within a nucleus that has to be overcome – can you think what that might be?

It's the repulsion of protons from each other in larger nuclei.

Once again it is the strong nuclear force which binds the nucleons within a nucleus together.

However, given that the range of the strong force is akin to the diameter of a single proton, how does this happen, since the arrangement of protons in nuclei can exceed this?

To accomplish this other particles come in to play to 'transmit' the strong force amongst nucleons, which we will consider later.

Let us now get some perspective on the relative sizes of atomic and subatomic structures.

We will start with the smallest particles and then work our way up in size:

The smallest particle size is the nominal 'diameter' of a quark.

This matches that of an electron (as far as we can determine) and this is 1×10^{18} m.

Of course, if we consider another model of the electron – the quantum model - then there is a non zero chance that an unobserved electron can stretch across the universe!

A small nucleus, which is taken here to be the diameter of a proton (assuming that a proton is spherical) is about one thousand times bigger and is 1×10^{-15} m.

A larger nucleus consisting of many more nucleons, which will be both protons and neutrons, is taken to be 1 x 10^{-14} m, or ten times that of a single nucleon.

Finally the diameter of a simple atom is some 1 x 10^{-10} m, which is one hundred million times the size of a single quark or electron!

Notice also that most of the volume within an atom is 'empty' space.

Put another way, if we scaled up the diameter of an electron or quark to say one metre, the equivalent diameter of an atom would be one hundred thousand kilometres!

Now here is a problem for you to consider; can you remember what the mass of two up quarks and a down quark is, these being the constituent quarks of a proton?

An up quark has a mass of about **4 MeV** and a down quark has a mass of about **8 MeV**.

Can you now remember the rest mass of a proton?

It is **938.257 MeV**.

What therefore is the percentage of the total proton mass that the quarks represent?

The answer is only about **1.7%**!!!!

How can this be possible?

The answer is of course that the 98% of 'missing' mass comes from energy in the form of kinetic and potential energies.

Actually, the strong force still remains the least well understood force of nature's four, and it is the strong force that is responsible for some 98% of the proton mass *(10.1)*.

But let's get back to the identical quarks that we spoke about in the beginning of this chapter – what is the force that repels them, and what is the force that attracts quarks of opposite colour charge?

This in another of nature's four fundamental forces:

2. The Electromagnetic Force

Every time you take a step along the corridor, why don't you just fall through the floor?

This question isn't as silly as you might think because the answer is not really that solid meets solid and they can't pass through each other – at least not on the quantum level!

Remember that atoms are mainly 'empty' space.

So why don't atoms just pass through each other?

The answer is that the electromagnetic repulsive force between atomic electrons keeps them apart and that is basically why we don't keep falling through the floor when we walk.

Or diagrammatically:

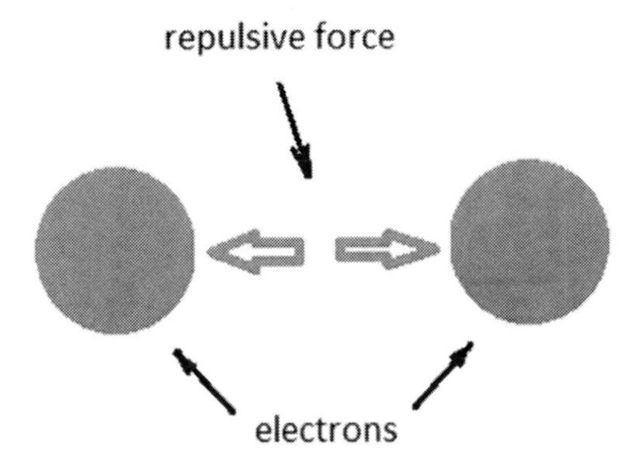

(figure 42)

The electromagnetic force only acts on charged particles, so that all neutral particles do not feel it.

This force is different from the strong nuclear force in a few ways. First of all it is weaker at similar ranges by about one thousand times.

Secondly, unlike the strong force which is a very short range force, the electromagnetic force is a long range force.

In fact, it extends to infinity, so that assuming that there were no other forces to interfere with it, an electron and a proton separated by an infinite distance would in theory begin to be attracted to each other if their separation were reduced by a small amount, although the size of the attraction would be infinitesimally small at such a range.

Unlike the strong force, the effect of the electromagnetic force decreases with separation according to the following inverse square law:

$$F_e = \frac{qQ}{4\pi\varepsilon_0 r^2}$$

where 'q' and 'Q' are the charges on two particles experiencing the force at a separation between centres of 'r', and 'ε_0'is the permittivity of free space.

But as charged particles get closer together, this law results in the electromagnetic force becoming very strong indeed at atomic and particularly subatomic ranges.

Look at how steeply the force rises with decreasing distance in the inverse square relationship below:

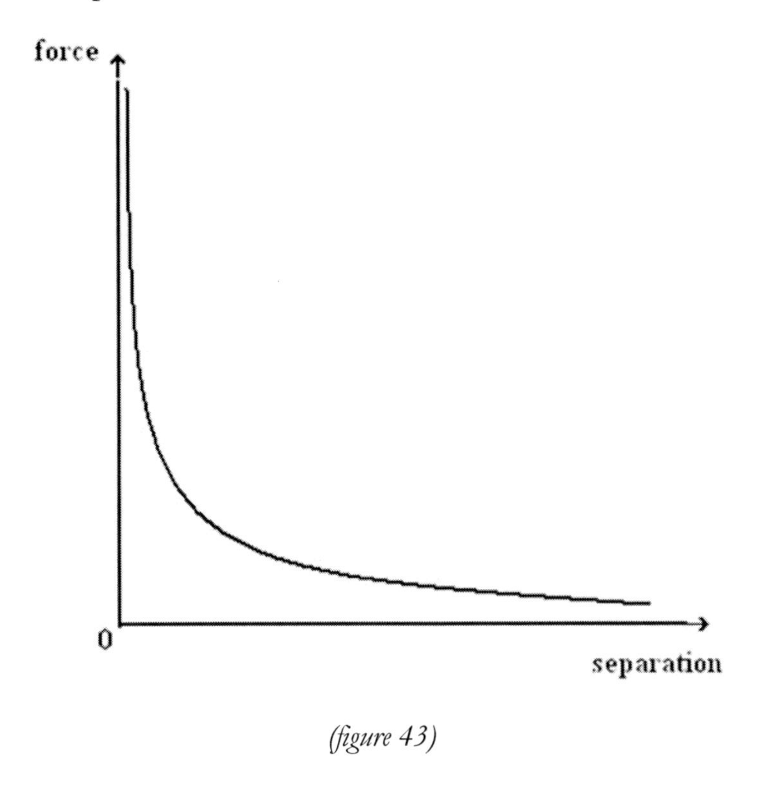

(figure 43)

What particles that we have encountered are influenced by the electromagnetic force?

All quarks and leptons come under its influence, but neutrinos are exempt since they are neutral.

Although neutrons contain quarks with colour charge, they are exempt as well because the fractional quark charges cancel out to zero.

3. The Weak Nuclear Force

Yes there is another nuclear force besides the strong force, and it is weaker than the strong force and hence its name, but don't be deceived by that – it is still a very strong force!

Just how strong or weak is it?

Remember that if we give the strong force an arbitrary strength of 'one', then the electromagnetic force is about one thousand times weaker.

The weak force, by a similar comparison, is only about 1×10^{-16} times as strong as the strong force, meaning that it is some ten million billion times weaker than the strong force and about ten thousand billion times weaker than the electromagnetic force at comparative ranges.

As you might guess, the weak nuclear force is a short range force, but its range is even less than the range of the strong force.

The range of the weak nuclear force is a mere 1×10^{-17} metre, or ten times the quark 'diameter'.

As its name also implies, its work is confined to the nucleus, so what exactly does it do there?

Its function is different in a sense from the other forces, which act to either separate or bind together various forms of matter.

The weak nuclear force has one purpose, and that is to control the emission of radiation from nuclei.

Whenever nuclear radiation is emitted from an unstable nucleus, the weak nuclear force is responsible for that emission.

For example, the alpha, beta and neutron radiations shown below can only happen if the weak force acts:

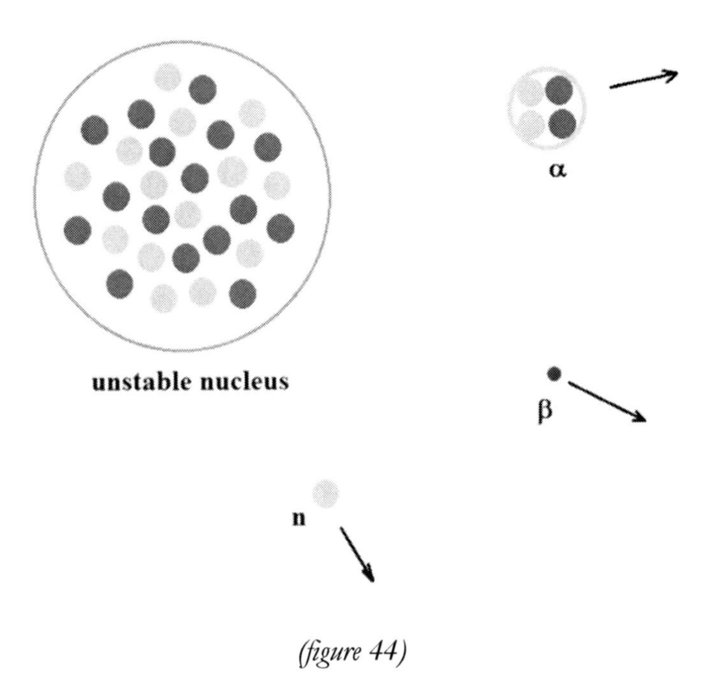

(figure 44)

Radioactive decay is a random process, but only as random as the weak force permits. Neither alpha nor beta radiation can be emitted from an unstable nucleus unless the weak nuclear force acts to facilitate that emission.

And its efficacy is not just limited to alpha and beta radiations – all particles that can be emitted from a nucleus only do so by the action of this force, so neutrons and neutrinos also come under its remit.

Just a relatively short time ago, it was thought that neutrinos had zero mass, but that now is not thought to be true.

This is because there seems no other explanation possible for the dearth of observed solar neutrinos detected on Earth.

Trillions more neutrinos are known to be emitted in the Sun's nuclear reactions than detection levels here indicate, but this can be explained if neutrinos oscillate between type during propagation between the Sun and Earth.

For this to happen, the decaying neutrino must have *mass.*

Before this was realised, the only one of nature's four fundamental forces that could have acted on a neutrino was the weak force, since the strong force is limited in its action to quarks and neutrinos are neutral so the electromagnetic force doesn't act on them.

However, as long as they do have mass, there is one more force that can act on them, albeit to an infinitesimally small level …..

4. Gravity

It was Sir Isaac Newton who seems to have been the first one to seriously think about what gravity did.

The story about the apple hitting him on the head may not have been true, but his insight into how gravity behaved was amazing.

He thought that if you fired a cannon ball ever faster and from a higher position, then eventually the Earth would fall away from the ball as it fell, due of course to its curvature.

There seemed no reason why, assuming the height and speed were sufficient, that the ball wouldn't continue its flight, all the time missing the Earth, and keep moving in a circle.

This of course is an *orbit*, and all orbiting spacecraft are simply falling, but missing the Earth or whatever planet they are orbiting, as it falls away from them.

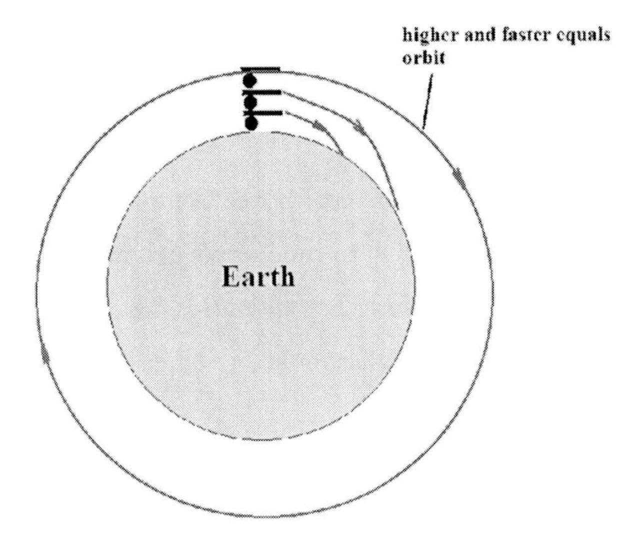

(figure 45)

What is amazing is how Newton realised that the Moon was actually in orbit around the Earth, and how he was able to derive the inverse square gravitational law that has served humankind so well with its space programmes!

Look at the picture of these astronauts enjoying the experience of a negative 'g' parabola in an aircraft.

(courtesy of NASA – http://grin.hq.nasa.gov/ABSTRACTS/GPN-2002-000039.html)
(figure 46)

They are weightless – right?

Actually they aren't – they have their full weight, but they can't feel it because they are in free fall.

In fact, astronauts in the International Space Station still have some 89% of their weight, because they are strongly in the Earth's gravitational field.

But again, they can't feel it because they are in free fall.

But how can the atoms in our bodies, or in any other mass, 'know' that they are in a gravity field?

Think about that question – it deserves some thought, because although being trapped in the Earth's gravity field is second nature to us, it doesn't explain why our atoms 'know' they are in that field!

The fundamental force of gravity is one with which we are all very familiar. Gravity, like the electromagnetic force, is a long range force and its effects stretch to infinity.

We all know that it is the force of gravity that causes planets to orbit the sun, like this:

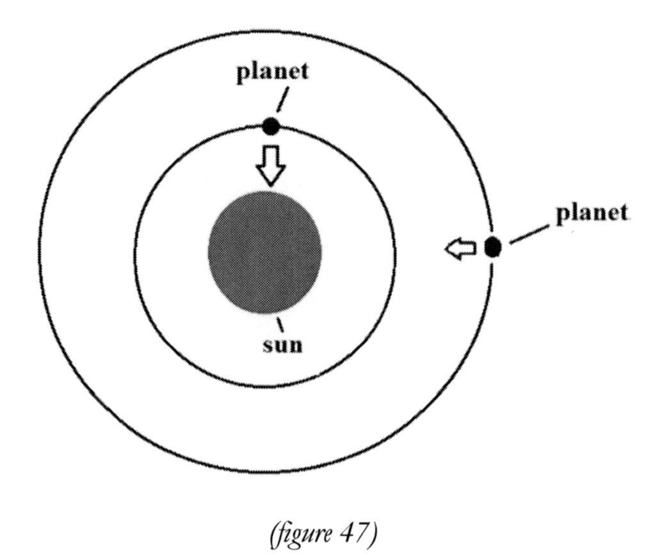

(figure 47)

But there are some surprising things about the gravitational force.

First of all, it acts on all matter that has mass.

You may not have realised it, but you exert gravity on this book and it exerts gravity on you, so that you are actually pulling each other towards you, but at a level that is totally negligible.

Imagine yourself somehow stranded in deep intergalactic space and away from the influence of all stellar and planetary gravity fields.

You are stranded in your spacesuit with another astronaut a few metres away from you who is also stranded.

As long as you are both motionless, then in theory you will gradually progress towards each other, pulled by your mutual gravity.

The second surprising thing about this force is that it also obeys an inverse square law of strength with separation, just like the electromagnetic force, and its strength is described by the following relationship – notice the identical format to that of the electromagnetic force.

The gravitational force of attraction between any two masses is given as:

$$F_g = \frac{GmM}{r^2}$$

where 'G' is the universal gravity constant, 'm' and 'M' are the two masses, and 'r' is the distance between centres of the masses.

Maybe you might want to muse on this question – why do these two fields operate on physical laws at all, and why are they identical in format, and how did the laws come about?

And remember, that although there is no theoretical reason why it should be so, these laws appear to hold everywhere where there is baryonic matter throughout this universe.

The third thing that is very puzzling about gravity, and which is of concern to not just a few cosmologists, is that gravity, in comparison to the other three forces, is *extremely weak.*

Now remembering that the weak force was about ten million billion times weaker than the strong force, you may be bemused by that.

But gravity is so comparatively weak that the strong, weak, and electromagnetic forces appear quite tightly bunched in comparison!

The gravity force is in the order of 1×10^{-41} times weaker than the strong force, which is about one hundred thousand billion billion billion billion times weaker!

The comparative weakness of gravity to the electromagnetic force is easily and intuitively demonstrated with a fridge magnet which will easily pick up a small steel object like a pin even though it is overcoming the gravity field of the entire Earth which is pulling it down!

Questions

1. Two large ships are docked together at a distance of 20 metres between centres.

If one ship has a mass of 20,000 tonnes and the other has a mass of 15,000 tonnes

(a) calculate the gravitational attraction between them

(b) would the ships have any realistic chance of moving together?

(take the universal gravity constant, 'G', to be 6.673×10^{-11} Nm^2kg^{-2}

and 1 tonne = 1000 kg)

2. Calculate by how many times the electromagnetic force of repulsion between two electrons is greater than the gravitational force of attraction between them at a separation of 1m.

Does this ratio vary with distance? Explain your answer.

(Take the universal gravity constant, 'G', to be 6.673×10^{-11} Nm^2kg^{-2}, the permittivity of free space, 'ε_0' to be 8.854×10^{-12} Fm^{-1}, the electronic charge to be 1.602×10^{-19} C, and the electron mass to be 9.110×10^{-31} kg).

CHAPTER 11: Bosons

All the fundamental particles that we have dealt with are fermions and we defined a fermion as a particle with half integer spin, such a 1/2, 3/2, etc, and all quarks and leptons have spin ½.

There is another class of particles called *bosons*, and these can be defined as particles that have integer spin, such as 0, 1, 2, etc, and which can also occupy the same quantum state if they have the same energy, whereas fermions can never occupy the same quantum state.

A *composite boson* is a particle that is composed of an even number of fermions, an example being mesons which have two fermions.

However, there are special categories of bosons that we will consider and these are *gauge bosons* and the *Higgs boson*, and these are the only ones of interest to us.

But there is another obvious question which first of all must follow on from the previous chapter and it is this; – how can an electron, which happens to be within the vicinity of another electron, be aware that it is being repelled?

After all, electrons don't have eyes, so how can they know?

This is just like asking the question as to how does the Earth know that the Sun is there pulling it with its gravity field?

The Earth can't see it, so how does it sense it?

Think about this carefully – imagine you are on the surface of the dwarf planet Pluto, which is about six billion kilometres from the Sun.

To you, there, the Sun would just be a pin point of light in the distance, and wouldn't be much brighter than any other star.

Pluto is cruising through local spacetime at a forward speed of nearly five kilometres per second.

Now think of a place you know about five kilometres away from you and imagine getting there in one second!

But actually Pluto is going nowhere fast, because that pin point of light in the distance has got it trapped.

Actually Pluto, and all the other (real) planets in our solar system, are orbiting the Sun, and in two hundred and forty eight years after Pluto passes any particular point in relation to the Sun, it ends up back there in the same spot.

The big question is this – how can Pluto possibly know that the Sun is exerting gravity on it?

After all, Pluto can neither see the Sun nor does it touch it.

Maybe we can find a clue if we start thinking about aircraft transponders.

A transponder is part of the operation of secondary surveillance radar.

A control tower equipped with radar facilities can be aware as to the whereabouts of an aircraft as long as that aircraft is fitted with a transponder.

The airfield will beam out a radar signal to a specific frequency that the aircraft transponder is set to.

The aircraft transponder, when it receives a signal from the field, automatically beams out a reply, and the two are in contact!

But remember that radar operates on particles – photons to be precise, and it is the photons that reach the aircraft that will cause its equipment to beam back more photons to the airfield.

This gives us a clue as to how two electrons in close vicinity of each other are made aware of each others presence.

The electrons will use *gauge bosons*, sometimes called *messenger bosons* or *messenger particles*, to probe the vicinity and 'seek out' similar bosons which might be beamed back to them, so gauge bosons are particles that *communicate forces* to other particles, and when this happens we say that the force is *mediated* by the gauge boson for that particular force.

Each of the four fundamental forces in nature is associated with its own gauge boson, which signals the presence of its field and which can indicate the presence of another source of the same force field, so let us now look at the bosons which are associated with each of the forces in turn.

1. The Strong Force Gauge Boson (Gluon)

The gauge boson which is associated with the strong nuclear force is called a *gluon.* Now a proton, for example, contains two up quarks and one down quark, and the question as to how each quark is aware of the presence of the strong force within its small range of operation is answered by the presence of gluons.

The implication of this is that we now see that there are additional particles within the limits of the proton diameter, namely gluons.

However, gauge bosons are considered to be *virtual* particles, because their existence is allowed for an extremely short time only, this being less than the Planck time.

So gluons 'inside' a proton might be imagined like this:

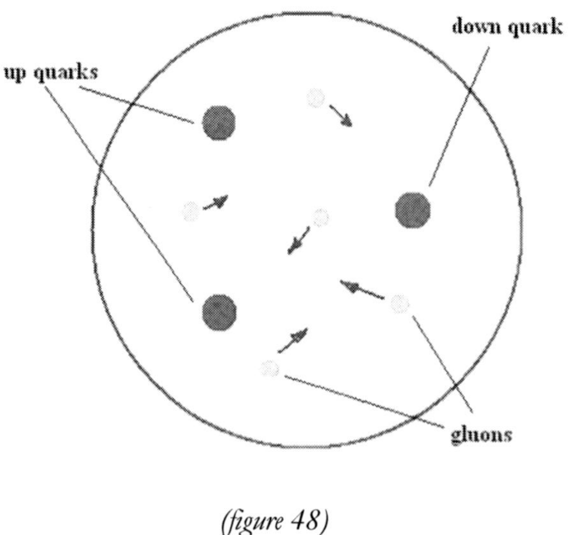

(figure 48)

What are the properties of gluons?

Gluons have spin 1, a mass of zero and they are neutral.

Their range is confined to the nuclear diameter of 1×10^{-15} metre.

This range is fine as far as communicating the presence of the strong force is concerned to the three quarks in a baryon, all of which are contained within the nuclear diameter, but don't forget that the strong force is also responsible for holding *nucleons themselves* together in a large nucleus, and remember that if this were not so, then nuclear protons would soon separate!

So how can gluons hold together nucleons, some of which can be separated considerably more than the nuclear diameter?

The answer is with the assistance of another particle.

Pion Mediation

The particle in question is one with which we are already familiar – it is the pion. Pions, of course, can be created in accelerators by, for example, the bombardment of beryllium or carbon nuclei with high energy protons, but pions are also found in nuclei.

We can think of this meson as something like a shuttle – transporting gluons amongst nucleons – something like this:

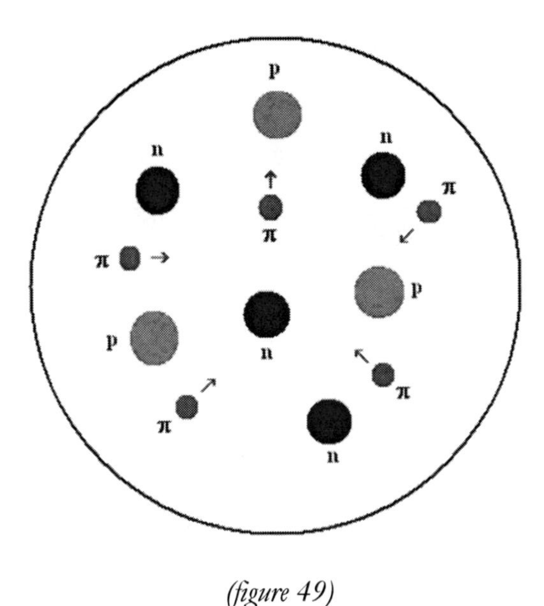

(figure 49)

Thus the sphere of influence of gluons extends beyond fundamental particles to combination particles as well.

Can you list the combination particles which you think come under the influence of the strong nuclear force?

They are quarks, mesons, baryons and atomic nuclei.

2. The Electromagnetic Force Gauge Boson (Photon)

Not only is it counterintuitive to think that an insensient electron orbiting a nucleus somehow knows that is in the range of the electromagnetic force from the proton or protons in the nucleus, but it is doubly strange that it is able to discern that it is in the vicinity of a positive charge!

The 'explanation' is, of course, that its knowledge comes by the mediation of the electromagnetic force boson.

This boson is a *virtual photon* which we might want to regard as similar to a *wave packet* like this ……

(figure 50)

Photons never interact with other photons, but they do interact with charged particles and we could be forgiven for asking just how, in the classical sense, this interaction takes place.

If we are thinking in terms of classical solutions, then we know that although photons have zero rest mass, they do have relativistic mass by virtue of their speed, and so they possess momentum.

Suppose we imagine that the repulsive electromagnetic force is mediated by the photon exchanging momentum with the like charged particle that it strikes – we can get a picture of that momentum transfer causing a motion of the target particle.

However, this analogy breaks down because this view does not explain attractive electromagnetic forces!

Of course, we cannot adopt a classical physical approach because we are dealing with quantum particles and our intuitive ideas fail at that level.

However, it shouldn't seem strange that a photon can be associated with electromagnetic forces, because remember that photons are packets of electromagnetic radiation that propagate via oscillating perpendicular electric and magnetic fields, the link between which is noted by the expression for the speed in vacuo and which is given by:

$$c = \sqrt{\frac{1}{\mu_o \varepsilon_o}}$$

Once again, this boson is virtual, because if it were not, then the laws of conservation of energy and momentum would be violated since such a boson must cease its existence in a very short time due to quantum uncertainty.

But we do know the attributes of virtual photons.

They have zero mass, spin 1, they are uncharged, and their range is infinity.

3. The Weak Force Gauge Boson (W^+, W^-, and Z^0)

When an unstable nucleus emits radiation the process is random.

That is to say that if we could shrink to atomic dimensions and we observed uranium atoms, we might see an atom emit an alpha particle, but we could stare at its neighbour for a long time because that atom might not emit radiation for a thousand years.

It isn't that the atom 'decides' to emit, it is that the weak force operates and instigates the decay.

The weak force is the only one of nature's four forces that deploys *more than* one boson.

In fact it deploys three, and these are the $\mathbf{W}^+$, the $\mathbf{W}^-$, and the $\mathbf{Z}^0$ bosons.

From their nomenclature we can see that two of the weak force bosons have attributes that no other bosons have, and that is that they are charged.

The $\mathbf{W}^+$ boson is positively charged and the $\mathbf{W}^-$ is negative, their charge magnitude being the electronic charge, but with the $\mathbf{Z}^0$ boson being neutral like all other bosons.

There is another striking difference between the weak force bosons and the bosons for the other forces, and that is that they have mass.

Actually it might be more appropriate to say that they have virtual mass since they are virtual particles and their mass is only possible as long they are virtual.

What sort of mass do they have?

Let's compare their masses with that of a proton, which has a mass of about 938 MeV, or 0.938 GeV.

The two **W** bosons both weigh in at some 86 times heavier than the proton, not forgetting that the proton mass itself is 1836 times heavier than an electron. So this puts the mass of the **W** bosons at 80.4 GeV or 80,000 MeV.

The neutral **Z** boson is even heavier with a mass of 91.2 GeV or 97 times the proton mass.

Maybe we can visualise these weak force bosons in relation to their size and charge something like this:

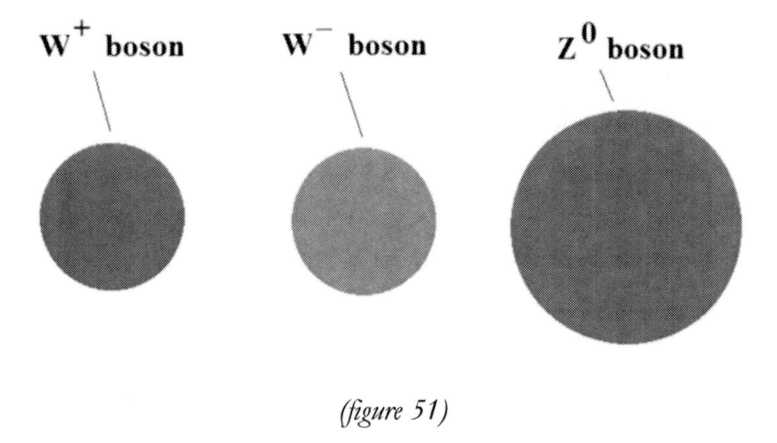

(figure 51)

There is a similarity with the other bosons that we have met, however, and that is that all three weak force bosons have spin 1.

The range of action of the weak force bosons is only one thousandth of the range of a gluon, being 1 x 10^{-18} metre, or a billion billionth of a metre.

The $\mathbf{W^+}$ and the $\mathbf{W^-}$ bosons are antiparticles and would therefore annihilate on contact, and the $\mathbf{Z^0}$ boson is its own antiparticle.

Since the two **W** bosons are charged, what are the consequences of their emission from another particle?

Clearly there must be charge conservation, so that the charge on the particle from which they come must be changed by one atomic charge unit in the opposite sense to the charge on the boson, or if this is inappropriate, other particles must be produced to carry the excess charge.

Neutron Decay

Let's look now at the operation of a **W**$^-$ boson.

The carbon 14 nucleus is unstable, and when it decays it emits a beta particle.

What is happening when that decay takes place?

Look at the decay equation:

$$^{14}_{6}\mathrm{C} \rightarrow {}^{14}_{7}\mathrm{N} + {}^{0}_{-1}\beta + \bar{\nu}_e$$

Clearly a neutron in the carbon nucleus has decayed into a proton to form the nitrogen nucleus, the atomic number of which is raised by one.

So what has happened to cause that?

Since the neutron consists of two down quarks and an up quark, and the proton consists of two up quarks and one down quark, can you deduce what must have happened?

Clearly a down quark in the neutron must have changed into an up quark to form the extra proton.

This, remember, can only happen by the mediation of a weak force boson, and the boson involved in that decay is the **W**$^-$ boson.

So the above equation can be reduced to:

$$\mathrm{d} \rightarrow \mathrm{u} + \beta + \bar{\nu}_e$$

In fact what has happened is that the down quark has emitted a **W**$^-$ boson, like this:

$$\mathrm{d} \rightarrow \mathrm{u} + \mathrm{W}^-$$

The **W** $^-$ boson then immediately decays into the beta particle and the antineutrino, like this:

$$\mathrm{W}^- \rightarrow \mathrm{e}^- + \bar{\nu}_e$$

Can you see how the charge is conserved by the emission of the **W**$^-$ boson?

The fractional charge difference between the down and up quarks is balanced by the $\mathbf{W}^-$ boson and in turn its charge is transferred to the electron.

Stellar Synthesis

Protons are extremely stable and no natural decay of a proton has yet been observed. However, if *energy is added*, then protons can be made to convert into neutrons.

An ideal place for the addition of energy would be the Sun's core, the temperature of which lies between ten and fifteen million degrees.

In dwarf stars like our Sun, nearly all of the helium is produced by a three stage process, the first of which happens when two protons interact.

Now two protons cannot form a nucleus alone, because such a combination is forbidden in this universe because the strong force cannot contain two protons alone. But two protons could be accommodated in a nucleus with the addition of neutrons.

The first stage in the process of helium production is the only one which concerns us, and in that stage two protons interact and one of them becomes a neutron, giving us deuterium.

First of all the protons must collide with very high kinetic energy to overcome the electrostatic repulsive force barrier – like this:

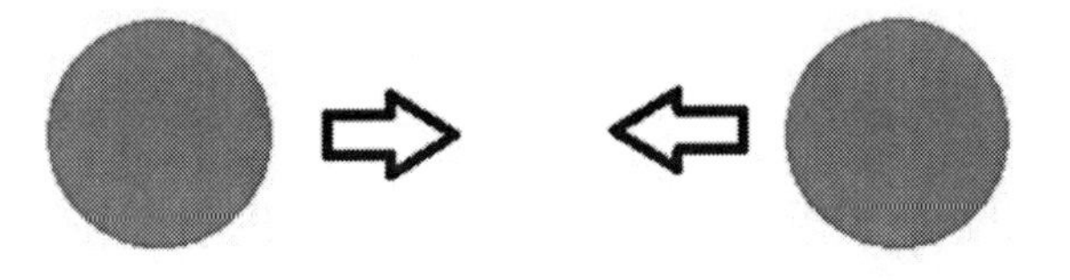

(figure 52)

If this happens then in order to facilitate the first stage in the fusion process, the weak nuclear force must act at the same time, in order to facilitate the conversion of one of the protons into a neutron.

This event is so rare that the average time for two protons in the Sun to undergo this process is longer than you might think – it is about one billion years!

Like this ....

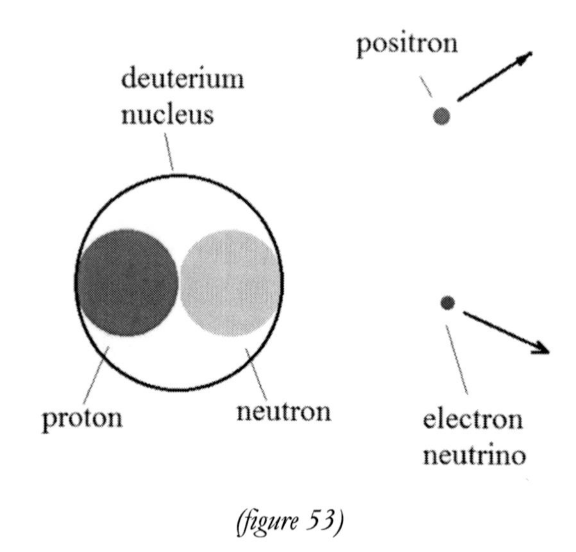

(figure 53)

This transformation can only happen by the mediation of a weak force boson, and the one involved is the $\mathbf{W^+}$. The equation describing this process is:

$$ {}^{1}_{1}\mathbf{H} + {}^{1}_{1}\mathbf{H} \rightarrow {}^{2}_{1}\mathbf{H} + {}^{0}_{1}\mathbf{e}^{+} + \boldsymbol{\nu}_{e} $$

In this case the opposite is happening in that an up quark in the proton is converting into a down quark to enable the neutron to form together with the production of a positron, the electron antimatter particle, and an electron neutrino.

Further, remember that this process can only happen *with the addition of energy.*

Once again, this cannot happen without the mediation of a weak force boson, which this time is the $\mathbf{W^+}$ boson, and once again charge must be conserved like this:

$$u \rightarrow d + e^+ + \nu_e$$

$$u \rightarrow d + W^+$$

The $\mathbf{W^+}$ boson then immediately decays into the positron and the neutrino, like this:

$$W^+ \rightarrow e^+ + \nu_e$$

That now leaves one weak force boson that we haven't yet discussed, this being the $\mathbf{Z^0}$ boson.

The exchange of $\mathbf{Z^0}$ bosons between particles is called a *neutral current* interaction and the charges on the particles involved are unchanged since the $\mathbf{Z^0}$ boson is neutral, but such exchanges involve a momentum transfer.

The $\mathbf{Z^0}$ boson can also be emitted in a particle antiparticle annihilation such as electron positron annihilation in which this boson can be emitted, and which then can decay into another particle antiparticle pair of another generation, such as a muon antimuon pair.

One strange characteristic of the weak force bosons is that they are capable of facilitating ***change of flavour!***

Note that the $\mathbf{W^+}$ and the $\mathbf{W^-}$ bosons can mediate the *change of flavour* of quarks and they can interact with both quarks and leptons.

Note that in neutron decay the $\mathbf{W^-}$ boson decays into an electron and an electron antineutrino, both of which are leptons and similarly the $\mathbf{W^+}$ boson decays into two leptons; the positron and the neutrino.

4. The Gravity Force Gauge Boson (graviton)

Our fourth force gauge boson is called the graviton and it can only act on particles that have mass.

It used to be thought that all gravity fields somehow permeated spacetime instantaneously such that if the Sun suddenly disappeared, the Earth would follow whatever tangential velocity it had at the time and lose its radial velocity instantly. However, that does not seem to be the case, because the Sun's gravity field is mediated by the action of gravity bosons, which have to propagate between the Sun and Earth through spacetime and which will take about eight minutes in our inertial reference frame, but which will take them zero time in their own reference frame since they, too, are massless.

This means that the Earth would continue to orbit the Sun's last known position for eight minutes (assuming it vanished from the Universe), even though the Sun didn't exist anymore!

The only problem we have at this time is that gravitons have not yet been observed. Nevertheless, we do know about their properties, and they are these; they have zero mass and zero charge, and its range is infinity, since gravity fields are infinite range fields.

But there is another difference between the graviton and the other bosons – it has spin 2 whereas all the others have spin 1.

We can summarise all of the particles that we have dealt with like this:

Fermions	**Bosons**
Leptons	**Gauge bosons**
Electron, muon, tauon, electron neutrino, muon neutrino, tau neutrino	Gluon, W^-, W^+, Z^0, photon, graviton
Quarks	**Mesons**
Up, down, charm, strange, top, bottom,	Kaon, pion, omega, etc
Baryons	
Proton, neutron, lambda, sigma, etc	

A Question of Mass

We have just said that the gravitational force boson, the graviton, can only act on particles that have mass.

Let's think about that for a moment – mass is made manifest in two ways – maybe. The two ways are gravitational mass and inertial mass.

I have just said 'maybe' because thus far in physicists' search for the truth, gravitational mass and inertial mass could be considered the same thing, but that isn't conclusive.

What are we talking about?

Well, gravitational mass is the property of matter, such as our bodies, that responds to gravitational fields, resulting in our being bound in the gravitational well that we are in on our planet.

But just think for a minute about this – if we were in deep space such that we were completely weightless, why should it be the case that we would still display the same reluctance to be accelerated that we do here on Earth?

After all, if there are no forces acting upon us in deep space, why should it take a force to accelerate us?

But it does, and that is inertial mass, and we know that we have inertial mass because we display reluctance to be accelerated horizontally on Earth, in which plane the Earth's gravity field doesn't act.

But to a very high degree of accuracy, both inertial mass and gravitational mass seem to be one and the same thing, but we can't be really sure of that.

So our question now is, how are massive particles endowed with mass?

And remember that possession of mass does not indicate possession of weight!

The Higgs Boson

In the late 1960s, the physicist Peter Higgs, and others, postulated a mechanism to enable matter to be endowed with mass, one symptom of which is inertia – the reluctance to be accelerated.

Although we could hold up a large truck with one finger in deep space because it is weightless, it would be a very different matter to try and push it!

The idea of a field in spacetime, now called a Higgs field, was proposed, through which could propagate another boson.

This field can be thought of as analogous to a viscous fluid such that a body passing through it would be hindered in its transit.

Any particle that can acquire mass which encounters the field will be influenced by it, more particularly by its mediating boson, the Higgs boson.

This boson is not a gauge boson in the sense that it communicates the presence of one of nature's four fundamental forces, but it is the boson that, were it to interact with certain particles, would endow them with mass.

The Higgs boson itself is endowed with mass within the field and its mediation causes other particles to acquire mass too.

It is hoped that the Higgs boson, if it exists, will be found in the LHC at CERN in Geneva, but even now we know some of its properties, and they are these:

It has no charge, and it has spin zero, and its mass is thought to lie approximately within the range of 114GeV to 160GeV.

The range of the Higgs boson is unknown, but is presumed to be short due to its mass.

We can now summarise the fundamental particles that make up the standard model of particle physics thus:

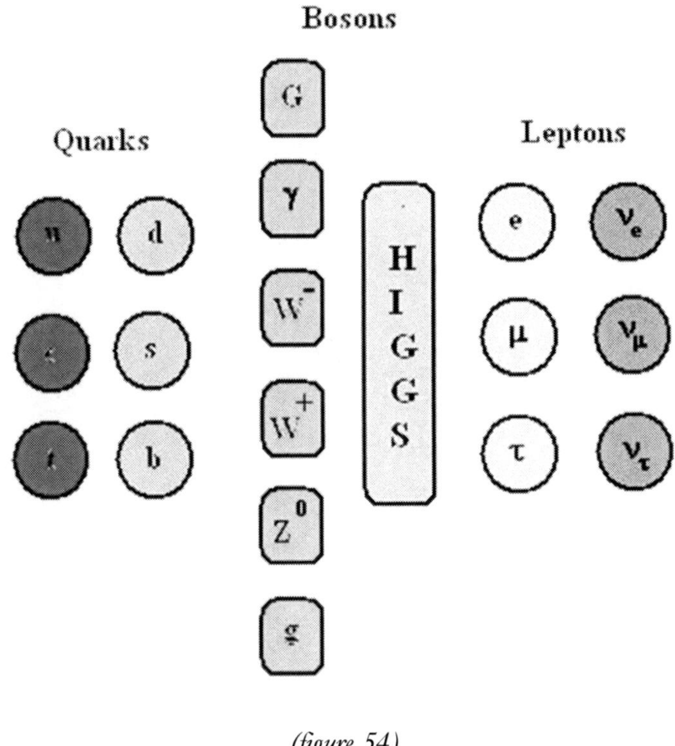

(figure 54)

Question

Given that the weak force bosons, the $\mathbf{W}^+$, the $\mathbf{W}^-$, and the $\mathbf{Z}^0$ are virtual, explain how their masses might be determined

CHAPTER 12: Feynman Diagrams

Boson Emission

Feynman diagrams provide an effective way of describing particle interactions in spacetime in diagrammatic form.

These diagrams have two axes, the vertical one representing progression of an interaction in time and the horizontal one representing the interaction in space. The propagation of a particle will be shown to change in the diagram when a boson is emitted, and there are a few ways to describe the emission of a boson.

One is by a sinusoidal line, which is sometimes used to describe the electromagnetic force boson, the photon, but is not exclusively used for that boson.

Real particles including antiparticles are always represented by a solid unbroken line, so the propagation of a particle pair which emits a photon could be described like this:

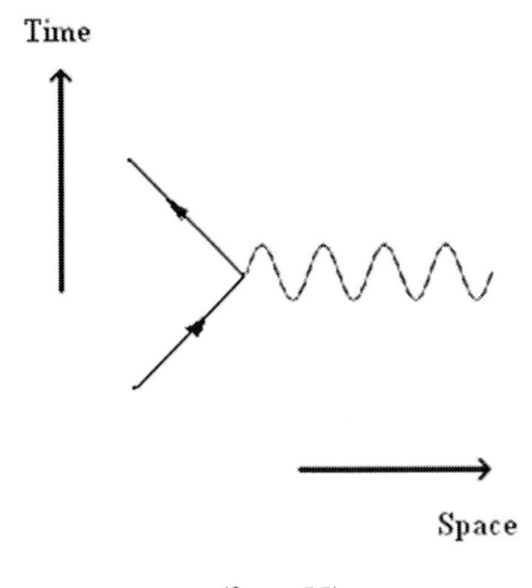

(figure 55)

This diagram would represent the annihilation of an electron positron pair, and in time ordered Feynman diagrams, matter particles are always directed from left to right, and antimatter particles from right to left.

Which particles are which in the above diagram?

The electron is represented by the bottom solid line and the positron by the top one.

Another way of depicting the emission of a boson is by a helical line.

This is sometimes used to depict the emission of a gluon, but not exclusively.

So particle propagation with gluon emission could be like this:

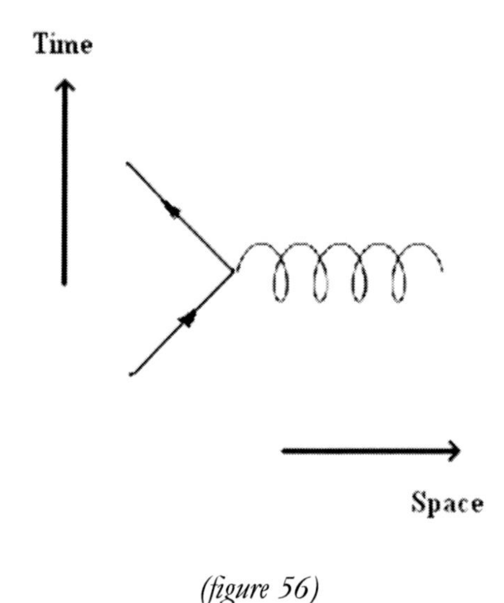

(figure 56)

A third way to depict boson emission is by a broken line, which is sometimes used for depiction of weak force gauge bosons, but not exclusively.

The following depicts a particle emitting one of the **W** or **Z** bosons:

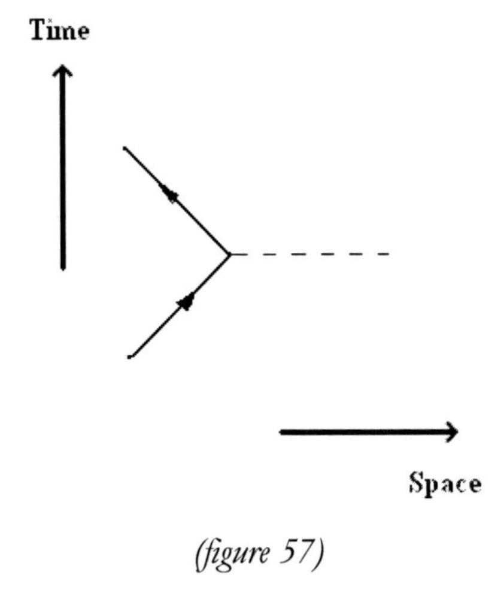

(figure 57)

If the interaction to be described were to be a high energy photon decaying into an electron positron pair, its depiction would be like this:

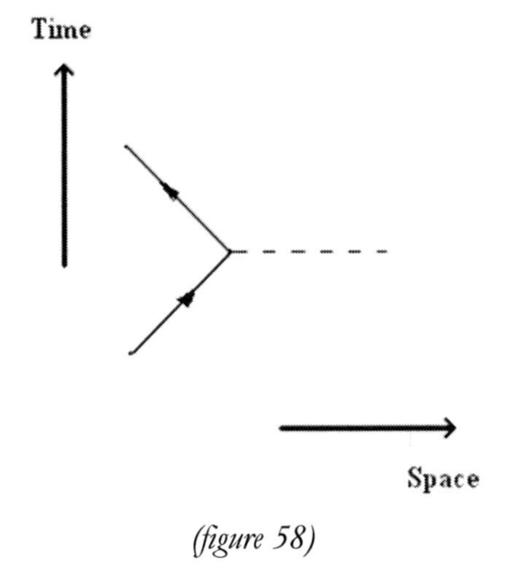

(figure 58)

In this book we will use the latter convention to describe *all* boson emissions.

Also it is not always necessary to draw time ordered diagrams and it is possible to draw the diagrams without time ordering, showing a 'picture' of the whole process, which we will be doing.

Now imagine two electrons which are converging on each other.

Each one will emit its 'radar' in the form of a photon, and this will result in a boson exchange between the two electrons.

Consequently the electrons, now 'aware' of each others presence in the vicinity, will change direction and begin to separate.

The following Feynman diagram depicts this:

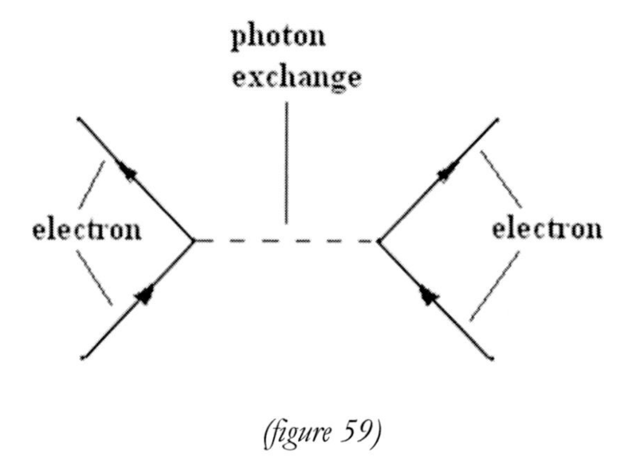

(figure 59)

Pion Exchange

Remember that in a nucleus the strong force bosons, the gluons, are mediated with the interaction of pions.

This too can be represented by a Feynman diagram, and the interaction between a proton and a neutron could be like this:

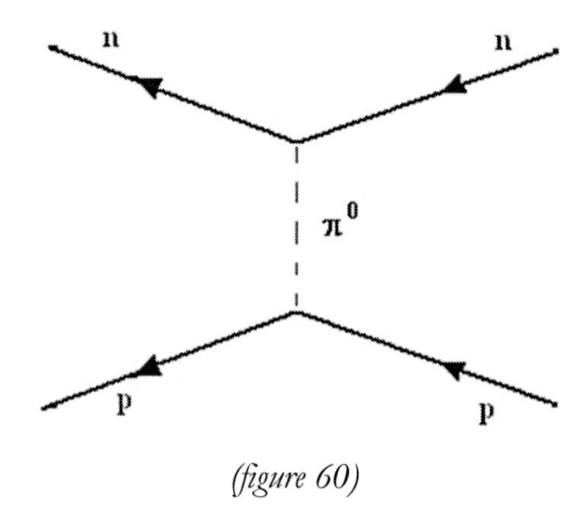

(figure 60)

Beta Decay

Remember also that when an embedded neutron in an unstable nucleus decays, a down quark in the neutron becomes an up quark by the mediation of a **W** $^-$ boson, which then decays into a beta particle and an electron antineutrino.

We can picture it happening something like that which is shown in the following diagram, but remember this diagram is *not* a Feynman diagram:

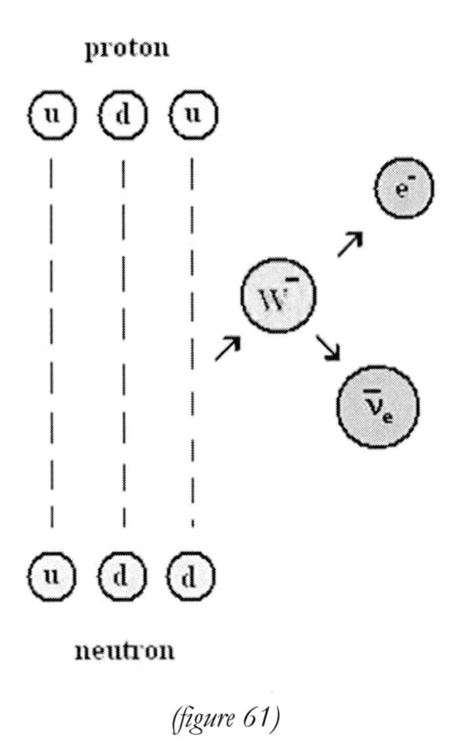

(figure 61)

Notice that the **W** $^-$ boson is emitted somewhere along the transition line between the down and up quark stages.

We can use this intuitive representation of what we think is happening as a guide to what the Feynman diagram for beta decay would be like.

The Feynman diagram for the process could be like this:

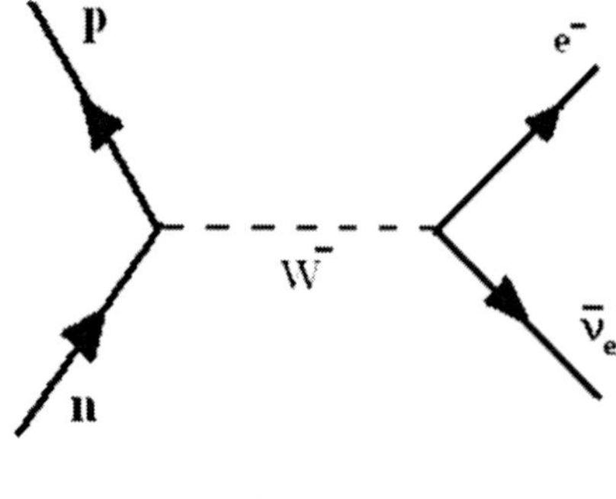

(figure 62)

More complicated diagrams can be drawn as well.

Do you remember muon decay?

A muon can decay into a muon neutrino, an electron and an electron antineutrino.

But what you didn't know thus far was that the original muon emitted a **W** $^-$ weak force boson, and in so doing changed the generation so that this boson subsequently decays into the electron and electron antineutrino.

This decay can be represented like this in a Feynman diagram:

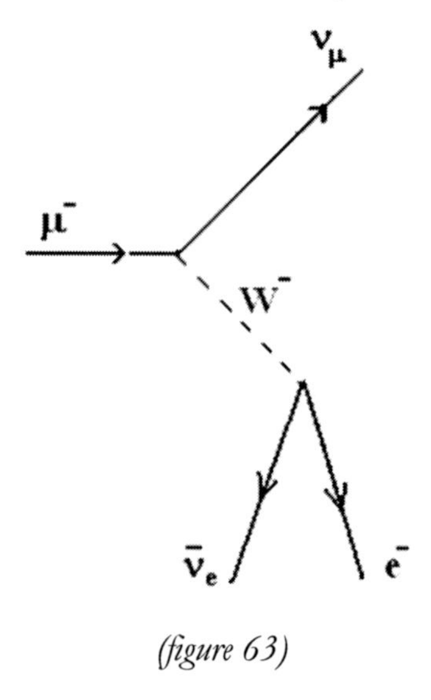

(figure 63)

Questions

1. Draw the Feynman diagram for a proton and an electron which are propagating in parallel within close proximity of each other.
2. Draw a Feynman diagram which could represent proton decay
3. Look in detail at the following Feynman diagram and describe what is represented there.

Your answer should include which particles are represented by the quark structure given:

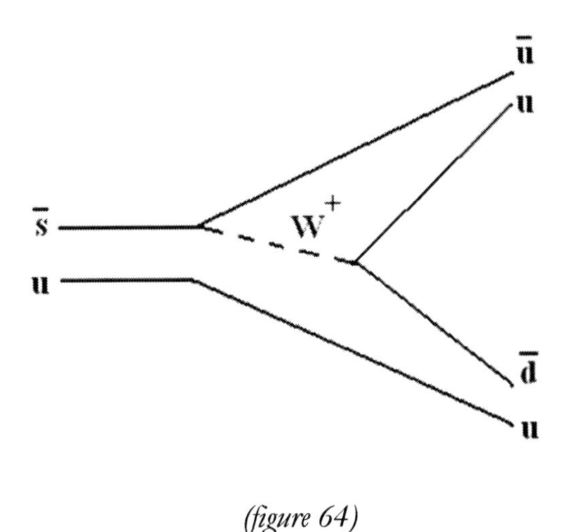

(figure 64)

CHAPTER 13: The Superforce, Puzzles and Solutions

There is plenty of evidence which points very strongly to the event which perhaps holds the most mystery and unanswered questions for many cosmologists, and that is the Big Bang.

This happened some 13.7 billion years ago, in which, possibly from nothing, an unmeasurable explosion of pure energy took place.

I have said from nothing, but that is if we do not count on other possible explanations which include some multiverse and oscillating universe theories in which our universe keeps on forever expanding and then contracting into a big crunch and then expanding again.

Before the Big Bang we assume that there was nothing, not even time, if we discount those theories.

And this thought leaves most cosmologists with a considerable degree of unease!

All the known laws of physics break down at the instant of the Big Bang because of the singularity problem in which all the mass energy of the universe appears to be concentrated in a singularity which has two attributes, these being infinite mass and zero size, which is of course impossible with our present understanding of the laws of classical physics and even quantum physics.

But to fully appreciate the beauty of what happened in the Big Bang, and how it seems that everything was somehow pre-determined, let's consider a few analogies first!

Before we start to consider what happened in the Big Bang, it might benefit us to think about a few things with which we are familiar – first of all an inflatable dinghy such as might be carried on a light aircraft.

It comes boxed in a very small volume and as soon as the activate cord is pulled behold a beautifully designed dinghy appears from the box, self inflating, and within seconds we have transport!

An even better example of this sort of thing could be an ordinary firework rocket. Before we set it off all we have is a small cylindrical piece of cardboard with a blue touch paper protruding from it.

But when we light the touch paper an array of things happen.

First we see the bright coloured sparks as the rocket climbs and then we might be likely to see an array of red, green, or some other coloured balls which appear and explode and possibly some other effects as well, all of them pre-programmed to come out of it!

(© www.bigfoto.com)
(figure 65)

The beginning of our universe was just like that in some ways – it is as if an amazing array of complex events fell out of the singularity, all of them appearing to be pre-programmed to give us the remarkable sophistication with which the events unfolded, and that in a very short space of time indeed!

The first thing to appear out of that singularity were very high energy photons, followed by the magnificent array of particles that we have been talking about, with the building blocks of the universe there in the melee, discreet in nature since other events would have to unfold, as if by magic, before they could amalgamate to form the stable baryonic matter that we are so used to.

Quarks for the only time in the universe's history (apart from a quark gluon plasma created in CERN) were discreet and could exist on their own, like leptons, without having to bind together with other quarks or antiquarks, together with gluons, forming the quark gluon plasma – but this state of affairs was only to last for a very short time.

It is as if all the laws and equations from which the universe would evolve were somehow programmed in, because what happened is far from some random chaotic miasma of events which had to undergo a series of random mutations before any sense was made of it all.

Amongst the first things that had to happen in our universe was the decoupling of all four of today's fundamental forces from the universe, one by one, and this apparently pre programmed sequence was all over in somewhere in the region of one thousand billionths of a second! …..

The Superforce

How many of nature's four forces were manifest at the instant of the universe's creation?

Actually at this stage in the history of the universe, nature's four forces were not thought to exist in the form they do today.

Instead, all the four forces were just one superforce, which could only be possible due to the extreme temperature.

But some unification of forces has already been successfully done by theory.

It started in 1864 when James Clerk Maxwell published his theory of unification of electricity and magnetism, and he successfully showed that electricity and magnetism were manifestations of the same forces.

Prior to that nobody had linked electricity with magnetism.

However, the first unification of two of the actual four forces came in the 1960s and 1970s when the electromagnetic force was unified theoretically with the weak nuclear force.

Experimental evidence in support of this unification came with the observation of weak neutral currents in 1973 and with the production and observation of the **W** and **Z** bosons in CERN in 1983.

It takes higher energy levels than occur naturally in our universe at this present time in its evolution for the fundamental forces to be unified, but at times close to the Big Bang conditions were right for this to happen.

But it also needs a theory that works!

The next step in unification would be to unify the electroweak force with the strong nuclear force, and this is referred to as *grand unification*, and the force which would then exist which unifies electromagnetism and the weak and strong nuclear forces is called the electronuclear force.

One holy grail in theoretical physics is to produce such a theory and this is called a Grand Unification Theory, or GUT.

Some grand unification theories exist, but these have only gained partial acceptance, but one thing that most of these predict is that the proton is inherently unstable, albeit with a lifetime of some 1×10^{33} to 1×10^{36} years.

Of course, since the Universe is only some 1.37×10^{10} years old, this has not been readily observed.

But remember that this figure for the lifetime of the proton is only an average, so it is possible that some may have decayed.

Some work has been done to try and observe such an event, because the lifetime given is statistical, but no decay has yet been detected.

Were proton decay to be detected, then further details of how grand unification could work might be glimpsed.

The last remaining unification after the electronuclear stage would be to somehow unite gravity with the other three forces, and this would then give us the single superforce which is thought to have existed just after the Big Bang.

Cosmologists talk about *spontaneous symmetry breaking* which occurred gradually as the temperature of the universe tumbled, and as symmetry was broken, so the forces and their boson masses fell out to be the way we see them now.

By way of an analogy, imagine four pencils bound together with rubber bands and standing on their ends like this:

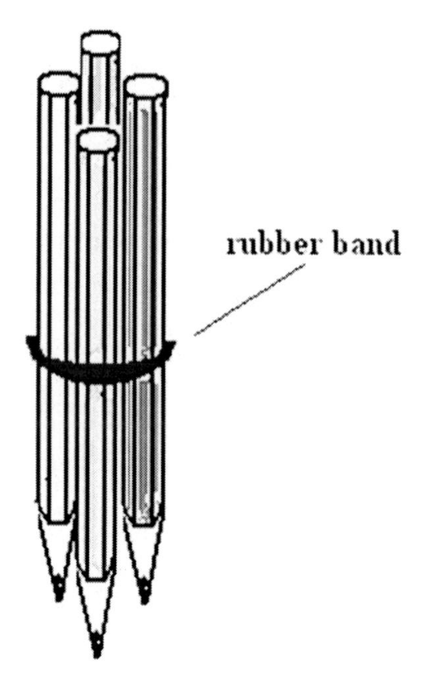

(figure 66)

If the rubber band holding all four pencils broke, then one pencil would fall out of symmetry, assuming there was another band holding the other three together.

That is an analogy of a high degree of symmetry being partially broken.

If the next rubber band then broke, another pencil would fall out of symmetry, assuming there was another band holding the last two together.

Similarly it is thought that as the extreme temperatures in the early universe began to fall, then one of what we now regard as nature's four fundamental forces would have decoupled from the original single superforce, the decoupled force being gravity, in the first instance.

The second force to decouple due to broken symmetry would have been the strong nuclear force, leaving the weak force and electromagnetism to decouple later, until eventually the four different forces were manifested as we see today at the temperatures that operate in the universe now.

However, one of the problems with a theory of everything is trying to restore the broken symmetry between the strong nuclear force and the electroweak force combination.

The problem arises with the strong force only acting on quarks and not on leptons. For unification to happen, leptons would need to convert to quarks and vice versa, which would, of course, violate conservation of baryon number in an interaction.

The higher that the ambient temperature is, the greater is the energy available for events to occur, so in the very early stages of the universe the energy available was very great indeed.

If the LHC at CERN can achieve energies of 1.4×10^{13} eV which is 14,000 GeV with head on collisions, we can soon see how little this in comparison with the energies available just after the Big Bang.

The Antimatter Puzzle

In the initial stages of the universe's existence the huge photon energy available ensured that an abundance of matter antimatter pairs were created, only to be instantly snuffed out again to be converted back into energy, and so the cycle continued, but only for a short time, because at around 1×10^{-35} second into the life of the universe, the strong force decoupled from the universe, together with bosons that we have not yet met.

As we know, the four fundamental forces that exist today are mediated by their own bosons.

Since, after gravity had decoupled from the universe, the electronuclear force predominated, it follows that it, too, had its own bosons.

These hypothetical bosons fall out of Grand Unification Theories, and they allow conservation of baryon number to be violated because they could couple quarks to leptons.

This is because they can theoretically allow a conversion between quarks and leptons by decay into an antiquark and an antilepton.

They are called the *massive X boson* and the *massive Y boson* and this is because their postulated mass could be of the order of 1×10^{18} MeV, or one million billion billion electron volts!

This is more than a million billion times the rest mass of a proton!

The massive X boson is hypothesised to compress baryonic quarks and to allow the mass of the proton to increase, thereby enabling the proton to become unstable and decay, but as the strong force decoupled from the universe, so these bosons disappeared from the universe too, and since our particle accelerators cannot reproduce the energies around in the GUT era, they have never been observed experimentally.

But there is another role which these bosons had and that is to play a part in the unequal balance of matter and antimatter seen in the universe today.

The excess of matter over antimatter seen today in the universe is the result of *baryogenesis* in the early universe and grand unified theories propose that this occurred by violation of baryon number conservation.

So when the GUT era was over, and the strong force decoupled, these bosons disappeared for ever from the universe as well, and thereafter something very puzzling occurred – for every one billion quark antiquark pairs that existed, one matter quark is thought to have remained without an equivalent antiquark.

As a result of this, the matter antimatter pairs eventually totally annihilated into photons as the temperature dropped, resulting in the huge outnumbering of photons to matter particles in the universe which is about one billion photons for each baryon, and is what we observe today.

But what cosmologists do not understand is why should there be the excess of matter over antimatter in the universe that we see today.

As far as we know, there does not appear to be an array of antimatter galaxies in some far corner of the cosmos, but the ultimate truth is that we do not know that for sure, although it does not seem likely.

One very good reason for this is that the electromagnetic force is so strong that in all likelihood attraction by now would have almost certainly occurred resulting in total annihilation, unless, of course, the antimatter equivalent of all the stable *neutral* atoms were somehow synthesised!

There are some clues as to how this could have come about however, and the first place to look would be at matter and antimatter particles.

We know that the only apparent difference between them is their charge, unless they are neutral, in which case they would likely be their own antiparticle.

But what if we found some other differences?

Could this explain some way out for antimatter from the universe?

First of all, if matter antimatter pairs are not to annihilate on immediate contact, there has to be a rapid separation of the pair in order to allow any asymmetry to gain precedence.

The universe was expanding so rapidly in the inflationary period that this could possibly have resulted in separation before annihilation.

But what sort of asymmetry could there possibly be?

One answer could lie in charge parity violation, sometimes shortened to CP violation. Charge conjugation involves the reversal of charge sign, which is a necessary requirement for a particle to become its antiparticle, whereas parity, or mirror symmetry, involves the reversal of all vector quantities on the particle concerned.

It was always assumed that the operation of charge parity was invariant, but then in 1964, to the surprise of many, it was discovered that for certain weak force interactions, the charge parity process was indeed violated.

Neutral **K** mesons can transform into their antiparticle by simply swapping their quarks for the other's antiquark and vice versa, so that the $(d\ \bar{s})$ configuration would become $(s\ \bar{d})$, thus delivering the $\bar{\mathbf{K}}$ antiquark.

However, over a large number of transformations, the probability of identical transformations from quark to antiquark and from antiquark to quark was non constant, showing charge parity violation in about one in five hundred decays.

This showed that CP violation was possible, and research is ongoing into the similar behaviour of **B** mesons, (which contain a bottom antiquark).

If this violation can lead to a theory of why the universe is dominated by matter, it is not without its problems, because the amount of matter over antimatter that might be generated by CP violation would only allow about one galaxy's worth of matter to exist, and not the 1×10^{11} baryonic matter galaxies that we know exist in the visible universe!

The Inflation Solution

Between the infinitesimally short times of 1×10^{-35} to 1×10^{-32} second, something remarkable seems to have happened.

We think the universe *inflated* at a very fast rate indeed.

The theory of inflation describes the universe increasing in volume at a phenomenal rate in this very short time interval.

The universe at this time was very much smaller than a proton, perhaps 1×10^{20} times *smaller*, but the universe, during the inflationary era, was able to *double its size* every 1×10^{-32} second.

What could have driven such a fast expansion?

We can only presume that it was a very strong version of the *dark energy* that is now driving galaxies apart.

After the inflation period, the size of the universe then would have been comparable perhaps to a beach ball!

Inflation solves a number of problems, and two of these are the flatness problem and the horizon problem.

The Flatness Problem

Spacetime itself is not necessarily rectilinear, and in fact there is only a slim chance that is perfectly rectilinear.

Our intuition would have us think that if we shone two parallel laser beams into deep space, then they would remain parallel.

That is not necessarily true!

It could be, for example, that many kilometres distant from the start of the laser propagation, the beams were slightly diverged on measurement.

Were this to be the case, it would mean that the very essence of the vacuum of space was curved negatively.

Given that the expansion of spacetime is actually accelerating, this might well prove to be the case.

On the other hand if measurement revealed that the beams had converged, and were a little closer together, the vacuum of space would be curved positively, like a sphere. This defies our intuition because we naturally assume that the beams would always remain parallel, if they were initially transmitted in parallel!

If this were the case, spacetime would indeed be flat.

But remember that in the early universe and before the inflation era, spacetime was, as far as we know, spherical, so that it might seem more logical to think of it as positively curved.

By way of analogy, imagine an ant crawling around the surface of a small balloon.

If the ant had sufficient intelligence it might soon realise that the surface it was crawling over was spherical, this being analogous to positive spacetime.

Like this .....

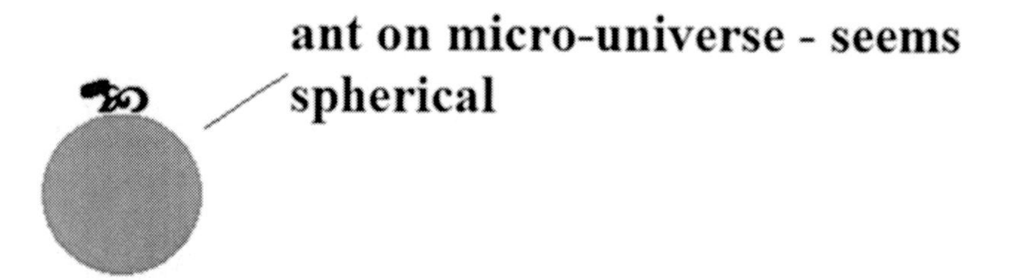

(figure 67)

However, if the balloon was suddenly inflated by a large amount, the ant could be forgiven for thinking that the surface upon which it was crawling was now flat.

Like this .....

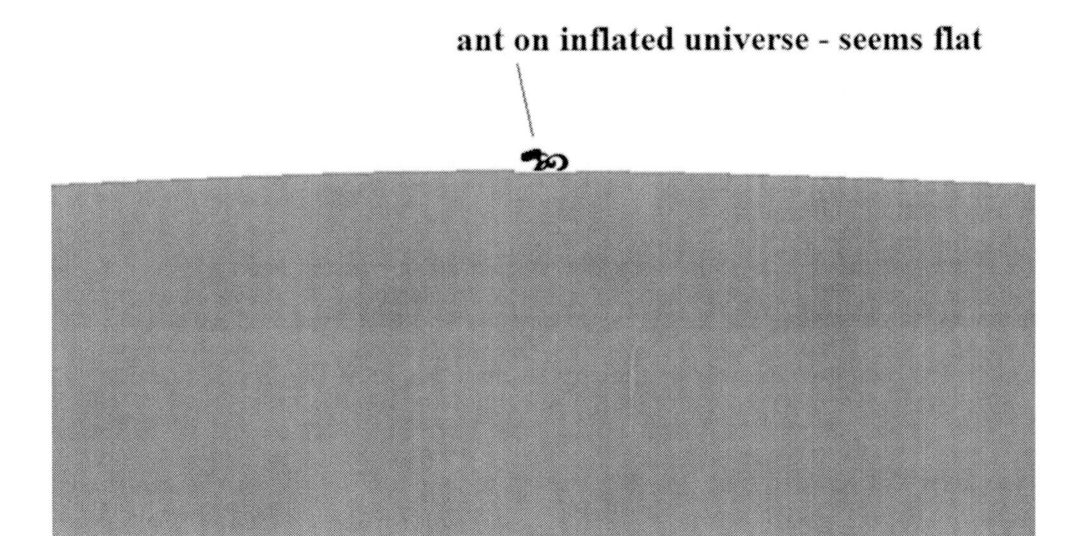

(figure 68)

As it happens, all our data seems to indicate that spacetime is something *approaching* flatness.

How can such a thing be?

Well inflation theory solves that problem in an analogous way to ant story.

This is true because the Universe has inflated so much that its curvature has been greatly reduced.

The Horizon Problem

If we look at distant galaxies, we see them as they were billions of years ago.

It is true that they are not like that now, because we are looking into the past, and the heavens that we see at the furthest extremities of our capabilities could only have been in existence for a relatively short time – so short, in fact, that light had only enough time to travel a distance related to just one degree of arc at that extremity compared with the three hundred and sixty degrees that our telescopes can scan!

That in itself is no surprise, but what is extraordinary, and has been a puzzle to cosmologists, is that no matter where our telescopes are pointed in the night sky, the universe is surprisingly homogeneous – in fact it is homogeneous to one part in one hundred thousand!

This is remarkable, because were the surface of our planet homogenous to the same degree, regarding the height of mountains and the depths of the oceans, the highest mountain would only be the approximate height of Big Ben in London!

Now given that thermal data could only have been exchanged by radiation then, how did thermal data get transmitted to regions outside of the range that light could have travelled in that time interval?

In other words, without some means of data exchange we would expect the universe to have displayed some greater degree of 'hot' and 'cold' regions.

It may be a poor analogy, but imagine a hand grenade going off.

One wouldn't expect to see equal amounts of metal debris in every direction.

On the contrary, some lines of sight would contain metal debris, whilst others wouldn't.

Equally, some regions would be hotter than others.

And if you don't see a problem, consider this; that if we wind the picture backwards, there is *never* a time, no matter how small the universe was in its early stages, that light had time to travel between regions, assuming, that is, that we *don't* have inflation theory to fall back on.

The problem is solved satisfactorily by inflation theory, since this offers an explanation as to how those disparate regions could once have been in contact.

If it were not for the cosmic inflation that we think took place, what do you think the diameter of the Universe would be now under its 'normal' slow expansion rates?

Were it not for inflation, the diameter of the universe at this present time, under 'normal' expansion rates, may only have been in the order of *one hundred millionth of a metre.*

So were it not for inflation, we could never have arrived here, unless, that is, the Universe had evolved microscopic humans!

Since 1998 evidence has emerged which indicates that the universe has been undergoing a much gentler rate of accelerated expansion instead of the decreasing rate that had been expected prior to this evidence, and which had been attributed to the inward attraction of gravity.

Pictorially, the expansion rate could be represented by the following figure:

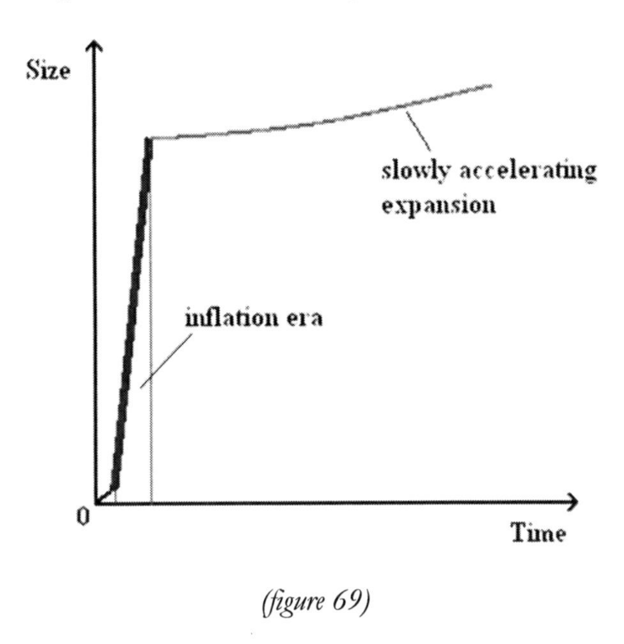

(figure 69)

CHAPTER 14: From Decoupling to Now

Let's work our way from the first moments and consider the temperatures and energies:

First of all at about 1×10^{-43} second after the Big Bang, the temperature of the universe was an inconceivable 1×10^{32} K, or one hundred thousand billion billion billion degrees.

Note that this was the first *real* time in the Universe, since the Planck time, which is 5.3912×10^{-44} s.

The Planck time is the shortest amount of time it is possible to have and no time before this could exist in the early Universe.

At this temperature gravity was the first force to decouple from the superforce as the energy available dropped below ten billion GeV, or 1×10^{19} GeV.

Note that this is in the order of a million billion times the maximum energy achievable in CERN at the present time and such energy will never be available in any particle accelerator so we will never be able to confirm this experimentally.

Up to this time all four forces existed as one superforce and this time is referred to as the Planck era in the history of the universe.

With gravity decoupled, the electronuclear combination continued to exist up to a time of 1×10^{-35} second when the temperature was some 1×10^{27} degrees with the energy available about to fall below 1×10^{14} GeV, or one hundred thousand billion GeV, at which time the strong force decoupled from the universe.

This is referred to as the *grand unification era.*

The electroweak force now existed alongside the decoupled gravity and strong nuclear forces, but this only lasted until a time into the universe of 1×10^{-12} second at which the temperature had fallen to 1 x 10^{15} degrees with an available energy just about to drop below one hundred GeV.

This short period is referred to as the *electroweak era.*

The decoupling from the universe of each of the forces in turn can be summarized like this:

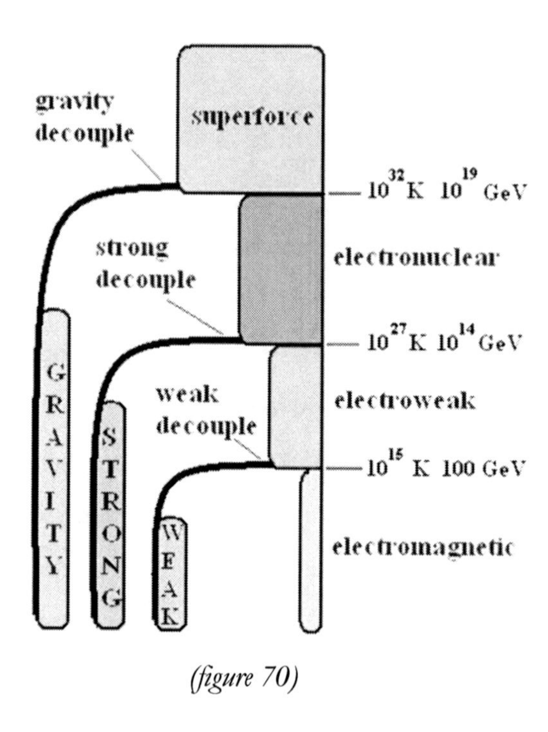

(figure 70)

Now how did all this fit in with the particle contents of the universe?

The universe then was of course completely unrecognizable compared with what we see today – first we had the dance of quarks and antiquarks, leptons and antileptons continually coming into and out of existence, swapping their existence with photons alternately, and this with some bosons present that aren't around today, and which were the mediation bosons for the combination forces that existed for that small fraction of a second, and with a decided lack of the ones with which we are now familiar, until the forces had decoupled from the universe, so that their very own gauge bosons could then accompany them.

Remember that all the baryonic matter, non baryonic matter and energy in our universe today, and we can only see some of it, was contained in that infinitesimally small volume immediately after the initial event.

It is a humbling thought that all the baryonic matter that we can see in every galaxy, of which there are one hundred thousand million, and every star, of which there are another one hundred thousand million in every galaxy, on average, and all the interstellar dust and hydrocarbon clouds only constitute some four percent of the total mass energy of our universe!

And remember that we are only talking about the *visible* Universe.

The distant galaxies that we can now see with the aid of the latest orbiting telescopes aren't there now.

The pictures they beam back are some 12 to 13 billion years old, so that we are firmly looking back into the distant past.

Those galaxies are now much further away – so far away that they can never be seen again by humans (unless there happens to be a big crunch starting sometime quite soon!).

Estimates of the actual diameter of the *whole* Universe now range somewhere between 90 and 160 billion light years!

Of course, no matter at all existed at the beginning as the total content of our universe was pure energy, but as we know, energy can be converted into mass and vice versa. With our knowledge of particle physics we might now be able to appreciate something of how our universe developed in the first few minutes, and most of it did happen in just a few minutes then, from the particle perspective, because after about a quarter of an hour nothing much happened for about three hundred and eighty thousand years!

It is the first quarter of an hour that we are interested in here.

We will examine what we think happened in very small time intervals just after the Big Bang happened.

At zero time the temperature of the universe was so high that it is unquantifiable. What filled the universe then?

The answer is very high energy photons and nothing else, but immediately, as the temperature began to fall, photons began to create matter antimatter pairs which in turn annihilated and converted back into photons.

This process happened continually.

These particles would have been lepton antilepton pairs such as electron positron pairs, and also quark antiquark pairs, and that process was set to continue for a very short time.

The universe at this time was in the Planck era and all the forces of nature were expressed in one superforce.

The first measurable time into the universe's existence is some 1×10^{-43} second

The reason that this is the first measurable time interval is because time also is quantised and no time interval can exist less than the Planck time.

The temperature fell very quickly to 1×10^{32} K, and at this stage gravity decoupled from the universe leaving the electronuclear force combination.

This is called the grand unification era, and matter antimatter particle pairs continued to come into and out of existence.

The universe was then composed of high energy photons, quarks, leptons, and gauge bosons plus equal numbers of matter and antimatter particles.

Remember that this includes the more exotic generations of leptons and quarks as well as the more stable variety.

No composite particles could exist because the temperature was so high that any such particles that might have formed would have been immediately ripped apart.

At a time of 1 x 10^{-35} second, the strong force decoupled from the universe, leaving the electroweak force combination to exist, the temperature at this time being 1 x 10^{27} degrees.

In this time interval, and continuing for time which stretches to about one millionth of a second into the history of the universe, although quarks and gluons existed, they couldn't yet form baryonic or mesonic structures because the temperature was too high and the quarks would be blasted apart.

As a result there existed what is called a *quark gluon plasma.*

The weak force had already decoupled from the universe at about 1 x 10^{15} degrees, but the quarks and gluons had to wait until the temperature fell to about 1 x 10^{12} degrees before they could fuse into baryonic and mesonic combination particles.

At that temperature of one thousand billion degrees the first protons and neutrons were able to form and this was the first stage of nucleosynthesis.

This is so because baryons now existed, although they had no chance at this stage of fusing into nuclei due to the high energy levels which would have instantly blasted any nuclei apart.

For *at least* the next three minutes, neutrons then began to exit the universe

The time is now 1 x 10^{-6} second, the weak and electromagnetic forces having already separated out leaving us with the four fundamental forces of nature that will exist from now on throughout the future universe.

After three minutes or so into the universe's creation the temperature is some 1 x 10^{9}, or one billion, degrees and the ratio of protons to neutrons in the universe is now 87% protons to 13% neutrons, and we will discuss the reasons for this in the next chapter.

This temperature is now cool enough to allow nuclei to fuse for the first time, and this is the second stage of nucleosynthesis.

The nuclei which form are predominantly helium and hydrogen, and the ratio *by mass* of these elemental nuclei is 74% hydrogen to 26% helium, and this ratio will remain throughout the future of the universe, apart from the negligible change due to helium synthesis in the later star fusion processes.

Besides the abundance of hydrogen and helium, very small quantities of deuterium and lithium nuclei also exist.

Remember also that photons, leptons, neutrinos and antineutrinos are abundant, but there is one attribute of the universe that is evident at this time which we do not observe today, and that is that the universe is *opaque*, so that no light can travel far.

Why is this?

It is because, although nuclei have fused, the temperature is still far too high for the positive nuclei to capture electrons and form atoms, the electrons having velocities in excess of the minimum required for electromagnetic forces between nuclei and the electrons to hold them.

In consequence, photons cannot travel far due to continual scattering off high velocity electrons.

This situation now lasts for a comparatively long time – perhaps three hundred and eighty thousand years or so, during which time nothing much happens in the universe except for its expansion and subsequent cooling.

After this time has expired, the temperature falls to around three thousand degrees, which is very cool in comparison with the former maelstrom, and finally the electrons are moving slowly enough for nuclei to capture them to form atoms.

This is called recombination, and for the first time, photons are free to travel through spacetime, unhindered by the miasma of free electrons.

The time when this took place is called the time of *last scattering*.

It is this picture of the universe that we see when we look at pictorial representations of the *cosmic microwave background radiation (CMBR)*, which has at our time in the universe suffered a wavelength stretch and is now at a temperature of about 2.7 K.

It is thought that the first stars lit up the dark universe after around four hundred million years with the first galaxies coming together at about one billion years into the life of the universe.

Some of these events can be depicted as shown in the diagram below which is not to scale:

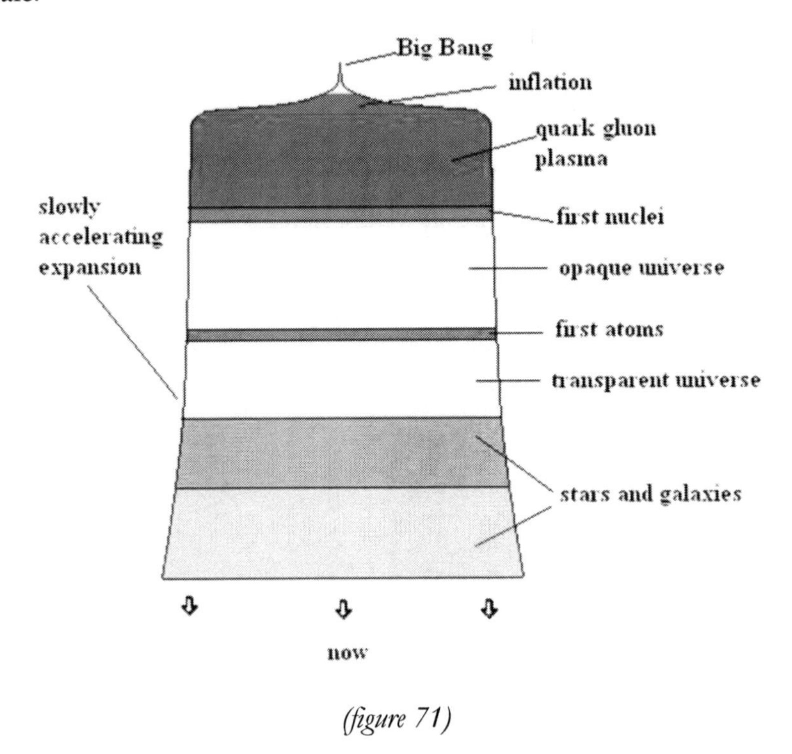

(figure 71)

From this time onwards, stars and then protogalaxies form, but it is not until some time later that there can be any planets.

Why is this?

The answer is that terrestrial planets such as Mercury, Venus, Earth and Mars can only be formed from the debris of a type 2 core collapse supernova and the universe had to wait until the first burn out of a star with a mass in excess of about nine solar masses for all the 92 elements in the periodic table to be produced, from which planets could be formed alongside the formation of a new non virgin star, such as our sun.

Our Sun is a *non virgin star* in so far as the original stars that were formed in the early Universe contained no heavy elements.

However, our Sun contains some 2% heavy elements, showing that it must have been formed from the debris of a type 2 core collapse supernova sometime maybe seven or eight billion years ago somewhere in our vicinity.

So some seven to eight billion years ago, such an event took place right here in our neighbourhood with the result that we are here.

Questions

1. Assuming that all the baryonic matter in the visible universe is the equivalent mass of 1×10^{78} protons, and that the proton rest mass is 1.673×10^{-27} kg, calculate how long the baryonic matter in the universe could keep a 100 MW power station running for assuming that it could all be converted into energy.

2. (a) If the diameter of the universe at the beginning of the inflationary era was 1×10^{-15} times smaller than that of a proton, what was its diameter?
(b) If the universe could double its diameter each 1×10^{-34} second, how many doublings could take place in the time for the inflationary era, which existed between 1×10^{-35} and 1×10^{-32} second?
(c) By how many times would the diameter of the universe have increased in that time?
(d) What would have been its diameter at the end of the period?

CHAPTER 15: The Neutron Story

Stage 1 Nucleosynthesis

As we learned in the last chapter, the weak force decoupled from the universe at a time of 1 x 10^{-12} second and a temperature of 1 x 10^{15} degrees, and at this stage all the basic building blocks of the universe were in place.

Then at about a millionth of a second into the universe's history, the first stage of pre programmed nucleosynthesis happened, in which baryons and antibaryons, and also mesons, began to fuse.

The main baryons that existed then were protons, neutrons, antiprotons and antineutrons, and it is assumed that equal numbers of protons and neutrons existed, and the same for their antimatter counterparts.

However because of the excess of matter over antimatter, a relatively small excess of protons and neutrons existed in comparison with their antiparticles.

Free Neutron Decoupling

But by now something of great significance was happening – the energy of photons was becoming insufficient to blast apart hadrons.

What would the consequence of this be?

It meant that for the first time matter baryons and antimatter baryons could hold together so that the antimatter baryons could at last annihilate with their antimatter pairs, leaving just the excess of matter hadrons in existence.

Since the neutrons in the universe were then not subject to being blasted apart by photons, they readily began to convert into protons which in turn converted back into neutrons by a process not now seen in nature, this being *neutrino interaction*. Remember that neutrinos today very rarely interact with matter, and billions from the Sun and other sources pass through our bodies and the Earth every second without any interaction.

But at the time we are looking at in the early universe, electron neutrinos could readily interact with baryons due to the presence of the abundant ambient energy at a temperature of thirty billion degrees (3×10^{10} degrees).

The interaction with neutrons resulted in proton synthesis at the expense of the neutron and vice versa.

At this temperature equal numbers of protons and neutrons initially existed, but when the temperature had dropped tenfold to some three billion degrees, the electron neutrino action ceased, but towards the end of this period the rate at which neutrons were destroyed began to exceed that at which protons were destroyed, with the result that the overall neutron numbers in the universe had decreased.

The consequence of this is obviously that the ratio of protons to neutrons in the universe had climbed above unity as neutron numbers diminished.

Stage 2 Nucleosynthesis

You will remember that free neutrons will decay in about 880 seconds, and this process now took over.

If this had been allowed to continue for any length of time then the universe would have been *without neutrons altogether* and as a consequence the periodic table of the elements wouldn't have existed and neither would we!

There would have been a proton universe and, with the values of the Universe's physical constants as they are now, the only element to exist would have been hydrogen!

But thankfully at just a few minutes into the life of the universe the temperature had fallen to about one billion degrees at which stage widespread nuclear fusion kicked in and helium four began to be produced, with nearly all of the remaining free neutrons forming half of the helium nuclei, and thence becoming stable for all time.

This was possible since at a temperature of some 9×10^8 K, deuterium became stable which allowed neutrons to very quickly combine to form deuterium and hence helium!

What is amazing here is this:

Somehow it seems that the rate of fall of temperature was pre-programmed to fall at exactly the right amount to enable nuclear fusion to take place at *just* the right time, when the 'correct' numbers of neutrons were left in the Universe.

What was 'correct' about these numbers?

The 'correctness' was that when the nucleosynthesis happened, the ratio of helium to hydrogen was 26% helium to 74% hydrogen, which is exactly what was needed to allow fusion to take place in dwarf stars like our own sun.

Without that mix of helium, there may well have been fusion in blue giants, but there would have been insufficient gravity in red, orange and yellow dwarf stars for fusion to kick in.

So what would be wrong with just blue giants fusing?

Blue giants have such short life spans that life would not have had time to evolve!

We need 'life friendly' planets around dwarf stars that will fuse hydrogen into helium, to give sufficient time for evolution to occur!

If the relevant constants in our universe had *not* been set at the specific levels that they were to enable fusion to take place at that time and temperature, then all the neutrons in the universe would have been lost!

This is called stage 2 nucleosynthesis.

At one time it was thought that the helium in the universe had all come from that produced by stellar nucleosynthesis, but it was soon realised that the abundance of helium in the universe was too great for this to be the case, and at about a quarter of an hour into the universe's history, the fusion process had just about been completed giving a ratio of hydrogen to helium nuclei in the universe of 74% to 26% by mass, with traces of deuterium, lithium and beryllium also present.

This ratio in the early Universe has been confirmed by observation of spectra from primordial interstellar dust in old galaxies.

And for a long time after that the universe would stay a very dark dark place – for some three hundred and eighty thousand years in fact!

CHAPTER 16: Dark Matter

The Laws

What is mathematically correct about our Solar System? ......

(Courtesy of NASA/JPL-Caltech)
(figure 72)

All the planets obey perfectly the Keplerian third law.

This law describes what the motion of the planets should be like.

This is according to the following law ......

$$v = \sqrt{\frac{GM}{R}}$$

Where **'v'** is the tangential speed of a planet,

'G' is the universal gravitation constant ($G = 6.673 \times 10^{-11}\ Nm^2kg^{-2}$),

'M' is the mass of the sun,

And **'R'** is the distance of a planet from the sun (usually quoted in Astronomical Units, AU)

The tangential speeds of the planets in our Solar System fit perfectly into this prediction, as is shown below ......

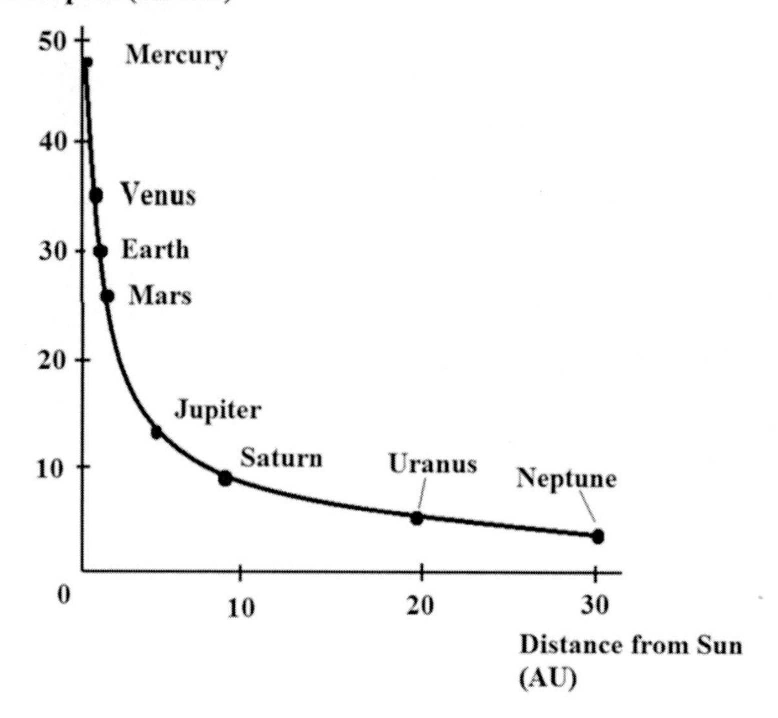

(figure 73)

Now look at the following picture of spiral galaxy NGC 4414 …….

(Courtesy of Hubble Heritage Team (AURA/STScI/NASA)
(figure 74)

Which way is it rotating with respect to the picture?

It is rotating anticlockwise.

Do we, on the face of it, have a mathematical problem with this, and all spiral galaxies?

The answer is that we *do.*

The problem is that the 'outliers', that is, the stars near the edges of the galaxies, do not obey the Keplerian third law with regard to their tangential speeds.

In fact, according to the mass of all the baryonic matter that we can see in the galaxy, that is, the masses of all the stars and interstellar dust, the stars near the outer limits of the galaxies should just fly off into space!

Why is this?

It is because there just does not seem to be enough mass in the galaxies to hold those stars in their positions!

Taking the Milky Way as an example, the radius of which is some 50,000 light years, the expected rotation curve for our galaxy would be according to the Keplerian decay curve, like this ......

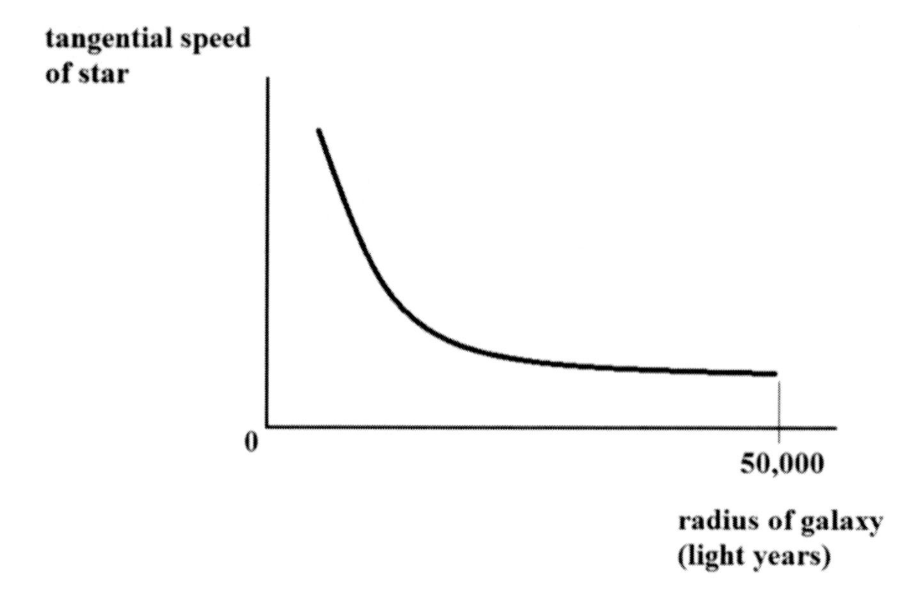

(figure 75)

However, it isn't!

It is more like this ....

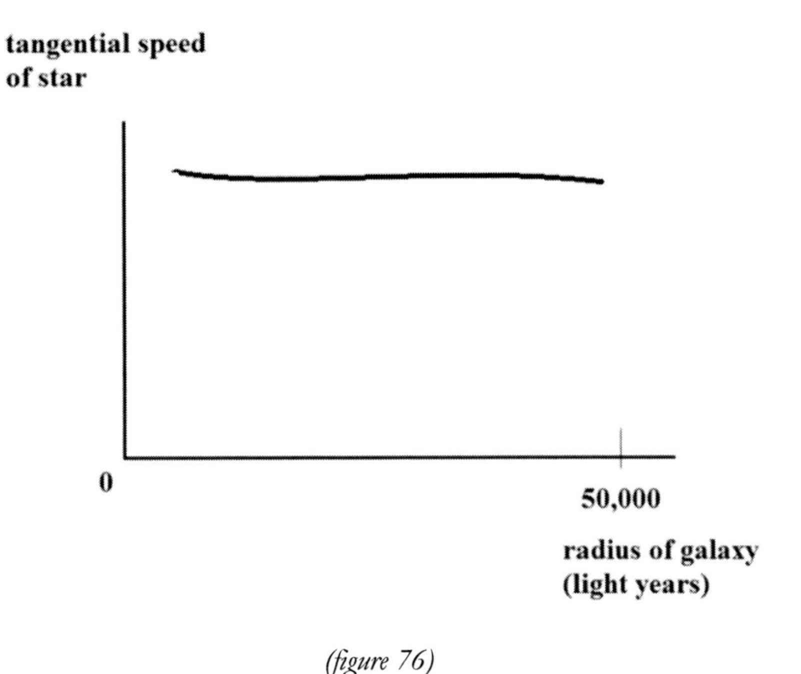

(figure 76)

What does this mean?

It means that there must be either something unseen *outside* the galaxy pushing inwards, or there must be something unseen *inside* the galaxy which is exerting a lot more gravity on the outliers than we knew about!

This unseen matter, if it exists, is called *dark matter* by cosmologists.

One of the properties of dark matter is that it does not react with photons, so therefore it can't be seen.

It is therefore invisible to all photon detectors, regardless of which electromagnetic frequency is being used in its detection.

Looking at the two curves together, will give us an idea of the difference in tangential speed that is attributable to dark matter in spiral galaxies ……

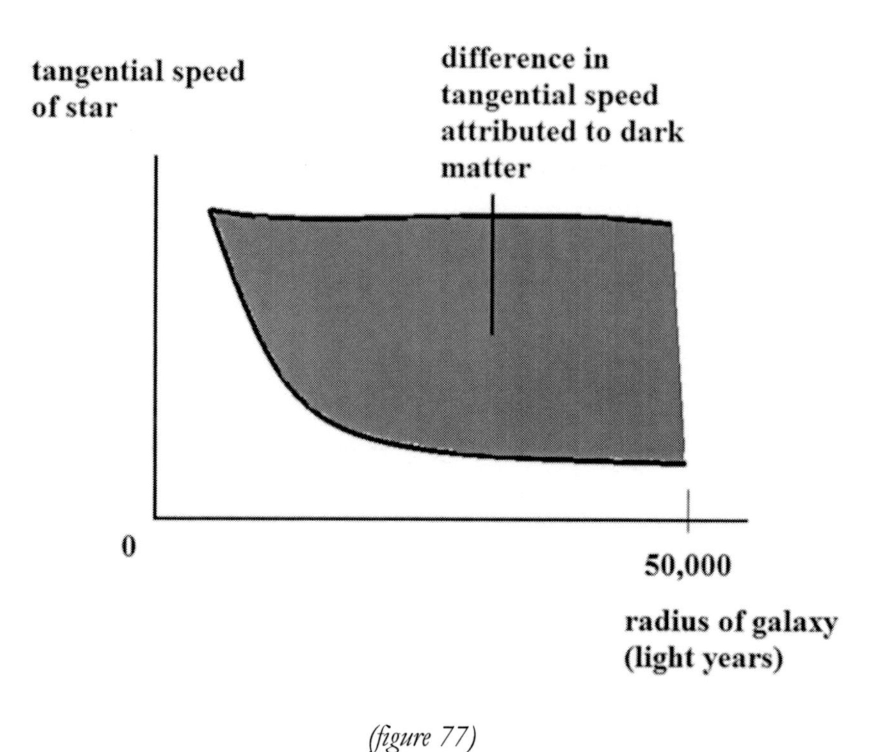

(figure 77)

What is dark matter?

The answer is that nobody knows at the moment.

However, there are some candidates for the post!

These include weakly interacting massive particles (WIMPS) which have never been observed, a proliferation of mini black holes, neutrinos and super symmetric particles.

Supersymmetry (or SUSY) postulates that every particle has a supermassive counterpart.

Examples are the neutralino, sleptons , and squarks, the 's' representing the supermassive counterpart to leptons and quarks respectively.

An example of one of the supermassive leptons is called a *selectron.*

However, no SUSY particles have been observed at the time of writing.

However, the recent discovery of very many more red dwarf stars in galaxies other than the Milky Way might well go a long way towards explaining where at least some of the dark matter might be found.

Red dwarf stars are inherently very dim and are therefore very hard to detect.

To summarize some of the proposed candidates that could possibly account for dark matter, here is a brief list ……

Brown dwarfs.

These are 'failed' stars which do not have quite enough mass to be able to sustain nuclear reactions in their cores.

Consequently, they are nigh on impossible to spot!

White dwarfs

These are the dead cores of yellow and orange dwarf stars which have a mass less than 1.4 solar masses.

Old ones will be losing their brightness as their temperature decreases, and so would be very hard to detect.

White dwarfs are very dense.

Neutron Stars

These very dense dead cores of larger stars are dark and extremely dense.

Black Holes

These are the densest objects in the Universe and are only detectable by the rotation rates of objects in close orbit to them.

Neutrinos

The problem with these as dark matter candidates is that they travel at near light speed and have very little mass.

As a result they could easily escape the immense gravity field of a galaxy.

Neutralinos

These are massive neutrinos, and are the proposed superpartners to neutrinos.

They could be a leading candidate for dark matter since they, due to their much larger mass, would be slower.

As yet no neutralino has been observed.

Axions

These are small mass, neutral particles.

Particle 'X'

Physicists have also proposed a new particle which is dubbed 'particle X'.

This particle would have a mass of some 1000GeV.

The interesting thing about this particle is that it is postulated to decay into either a neutron or into two hypothesised particles, the **Y** particle and the **Φ** particle.

These hypothesised particles would have a proposed mass of about 2 – 3GeV.

Research physicists in the Brookhaven National Laboratory in New York and the TRIUMF laboratory in Vancouver have proposed the idea.

The antiparticles to the **Y** and the **Φ** particle, the $\bar{\mathbf{Y}}$ and the $\bar{\mathbf{\Phi}}$ would make protons decay.

Hence evidence for this could come from existing experiments which can look for proton decay. *(16.1)*

Modified Newtonian Dynamics (MOND)

One alternative explanation for dark matter is *Modified Newtonian Dynamics* (or MOND).

This idea postulates that in the event where gravity is very weak, as in the outer reaches of a spiral galaxy, then Newton's laws might need modification, and that gravity is actually stronger there than we currently predict.

However, the idea is thought to be inconsistent with the conservation of angular momentum.

Actually it isn't very difficult to calculate the amount of dark matter that there appears to be in our own galaxy, the Milky Way.

Using Newton's law of gravitation and equating it to the centripetal force on an outlying star will help you with the following question which will enable you to make an approximation of the dark matter in our galaxy .....

Question

Data: **Universal Gravity Constant, G = 6.673×10^{-11} N m^2 kg^{-2}**

Speed of light in vacuo = 3.0×10^8 ms^{-1}

Average mass of galactic star is 7.1963×10^{26} tonnes

Number of stars in Milky Way = 2×10^{11}

Diameter of Milky Way = 100,000 light years

Periodic time of Milky Way = 250,000,000 years

Using the data provided, and considering the Milky Way galaxy

(a) Calculate the radius of the Milky Way in metres

(b) Calculate the baryonic mass of the Milky Way

(c) By equating gravitational force between an outlying star and the galaxy with the centripetal force on that star, calculate the maximum tangential speed that an outlying star could have (assume the mass of the galaxy to be centralised)

(d) Using the expression $v = \omega r$, calculate the actual tangential speed of the outlying star

(e) How much greater is the actual tangential speed compared with the maximum that could be achieved by gravitational attraction of the galaxy?

(f) Now calculate the ***actual mass of the Milky Way*** by using the same equations as in part (c), but using the actual tangential speed observed for the outlying star

(g) How much greater is this than the baryonic mass of the galaxy?

(h) Therefore how much dark matter is present in the galaxy compared with its baryonic mass?

CHAPTER 17: Dark Energy

Before Hubble

When do you think that we first knew that there were other galaxies?

Actually it wasn't until well into the twentieth century, and the person who first discovered that was the great astronomer Edwin Hubble in the 1920s.

Before that, although galaxies had been seen on telescopes, they were presumed to be part of the Milky Way galaxy, which, of course, was thought to be the Universe.

So certain were cosmologists then that the universe was static, that even though Einstein's general relativity field equations predicted an expanding universe, he went to the trouble of inserting a *cosmological constant* into the equations to make them fit the view of a static universe.

When Hubble discovered the *red shift* that proved that the universe was expanding, Einstein called his cosmological constant his biggest mistake!

However, that constant may now be gaining acceptance once more!

The Red Shift

The sound from a passing fire engine with its sirens going adopts a lower pitch as it recedes from us at speed.

This means that it therefore must suffer an increase in wavelength.

This is quite obvious since it is receding at speed, and therefore a greater distance from the observer is defined between each sound emission.

Although in astronomical observations, the speeds involved are much greater than those of a fire engine, the equivalent was observed by Edwin Hubble with electromagnetic waves.

In other words, he became aware, in his observations, of a shift in absorption spectrum lines of gasses and dust around distant objects.

This shift was always towards the *red end* of the spectrum.

In other words, the light from distant galaxies must have suffered *an increase in wavelength*, $\Delta\lambda$.

The *red shift,* **z**, is defined as the ratio of the increase in wavelength, $\Delta\lambda$, to the expected wavelength, λ.

So that ......

$$Z = \frac{\Delta\lambda}{\lambda}$$

This could only be because they must have attained a great speed relative to the Earth.

And what is more, Hubble discovered that there was a linear relationship between the distance away from us of a distant galaxy, ***d***, to the red shift, **z**!

In other words, the further away a distant galaxy was, the faster was its speed of recession.

Look at the diagram below.

The bottom line gives us the laboratory reference with no motion applicable.

The absorption spectra on the second line from the bottom are those which might be obtained from a nearby star.

Those on the next line up could be from a nearby *galaxy*

Those on the next line up could be from a more distant galaxy, whilst those on the top line could be from a very distant galaxy.

Notice how the absorption lines *shift* towards the red end of the spectrum, the further away their source is!

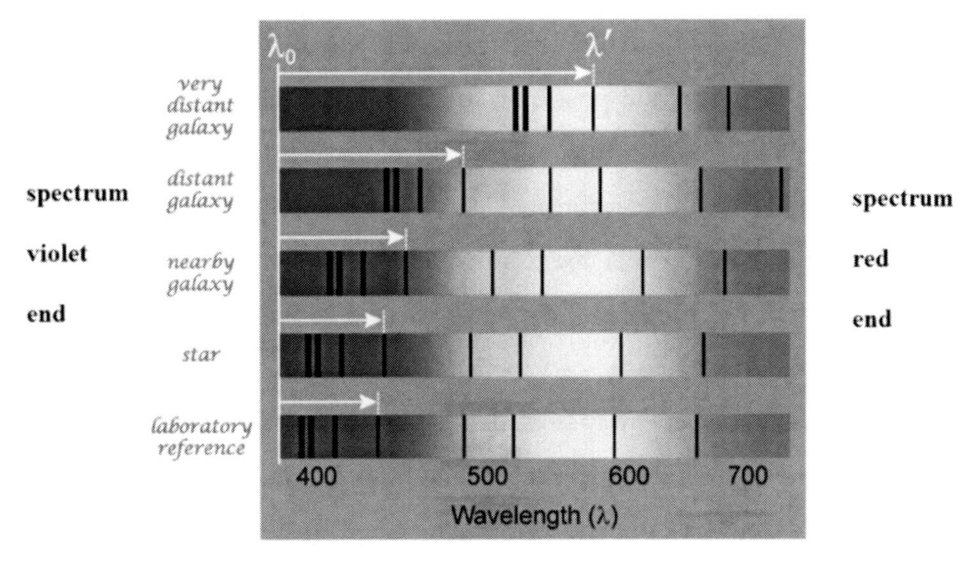

(Courtesy of NASA/JPL-Caltech)
(figure 78)

Further, Edwin Hubble noted that the linear relationship between the speed of recession of a galaxy and its distance away was like this ....

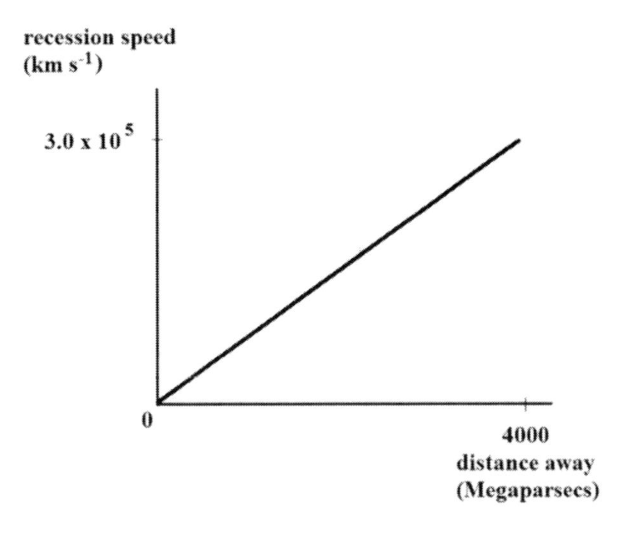

(figure 79)

Until near the end of the twentieth century, most cosmologists believed that all the gravity in the universe would slow down this expansion of spacetime, and reduce the recessional speed of distant galaxies, and could possibly stop their recession altogether, with the possibility of the idea of an eventual big crunch taking place sometime in the future.

However, something quite surprising was discovered in the late 1990s…

Saul Permutter and others in the team which ran the *Supernova Cosmology Project* at Lawrence Berkeley National Laboratory in the U.S.A., and others in the *High – z Supernova Search Team*, also in the U.S.A., found that the distance that certain *Standard Candles* were away from us was greater than was predicted!

The Supernova Cosmology Project and the High-z Supernova Project developed methods to seek out and then follow type 1(a) supernova explosions.

The groups scanned patches of sky to find this type of supernova and then monitored their evolution to record measurements of spectra.

In 1998 the Supernova Cosmology Project reported that supernovae with a particular red shift were dimmer than expected.

These results were soon confirmed by the High-z Supernova team.

This could have meant only one thing – that the expansion of the universe was not slowing down at all - but was actually accelerating.

Type 1(a) Supernovae

One must be careful not to associate type 1(a) supernovae with type 2 core collapse supernovae, from which all the naturally occurring heavy elements, with elemental masses greater than that of iron (which is the heaviest element that can be synthesised in stars) are created.

Type 1(a) supernovae occur in binary systems.

In this case, a carbon – oxygen white dwarf dead stellar core which happens to be less than a certain limit, plus possibly a red giant star.

This limit is called the Chandrasekhar limit.

The Chandrasekhar mass limit puts a value on the maximum mass that an astral body can have before electron degeneracy pressure is exceeded in terms of the Pauli Exclusion Principle.

This limit is 1.4 solar masses, or 2.85 x 10^{30} kg.

If a carbon – oxygen white dwarf remnant, which has mass less than the Chandrasekhar limit, is in binary with, for example, a red giant star, the intense gravity of the white dwarf can accrete matter from the red giant.

The artist's impression below shows a white dwarf core to the bottom left accreting mass from a binary companion ……

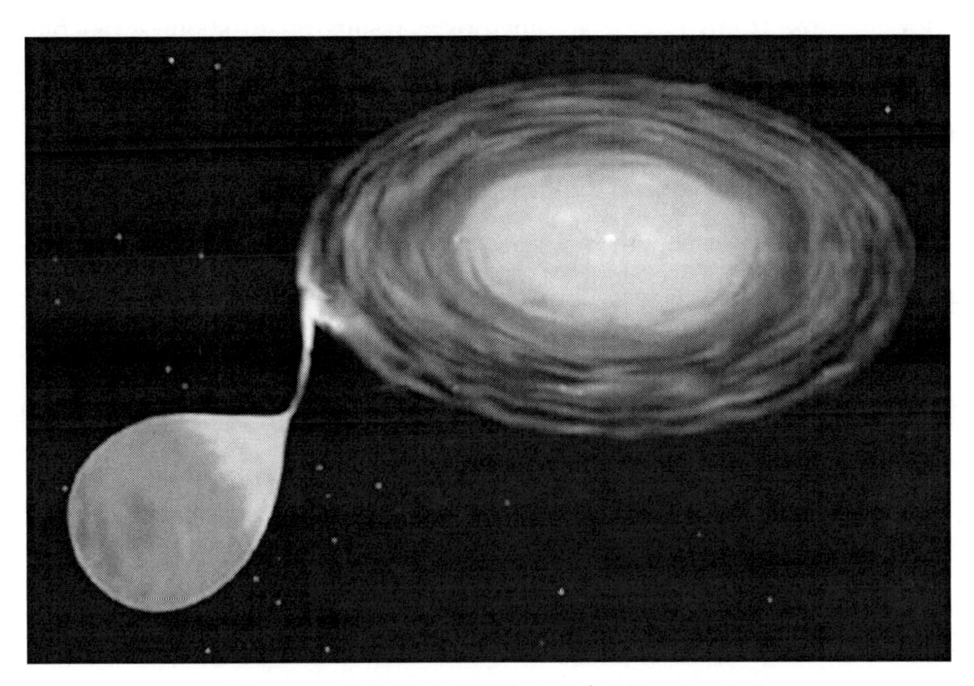

(Courtesy of NASA, HST Artist's Visualization)
(figure 80)

Once the Chandrasekhar limit has been achieved due to the mass accretion though, collapse begins to occur and within a few seconds it is believed that a substantial amount of the dead core's matter undergoes nuclear fusion resulting in an explosion.

This explosion is of the order of some *five billion* times the brightness of the Sun, and furthermore we believe that the brightness *is always the same* or with little variation.

This being the case, a type 1(a) supernovae can be used as a *standard candle*, which, according to the observed brightness of the explosion on Earth, can be used to determine how far away the event was.

So astronomers expected to be able to predict the distance away of this type of supernova according the brightness observed.

However, the teams looked at supernovae results up to about seven billion years ago and, to their surprise, found that the red shift and brightness did not correlate according to what was expected.

In fact, the supernovae were dimmer for a given red shift.

This meant that for supernovae at a given distance, the red shift was not as large as had been expected according to the standard model.

This could only have meant that in the distant past, the universe was not expanding as fast as it is now, and hence the wavelength of the light seen was not red shifted as much as it should have been.

This could only mean that the universe was expanding at a slower rate in the distant past compared with now.

In other words, the speed of expansion of spacetime has *increased*!

The diagram below shows how the results indicated a smaller red shift observed for a given distance for red shifts of about 0.2 upwards ……

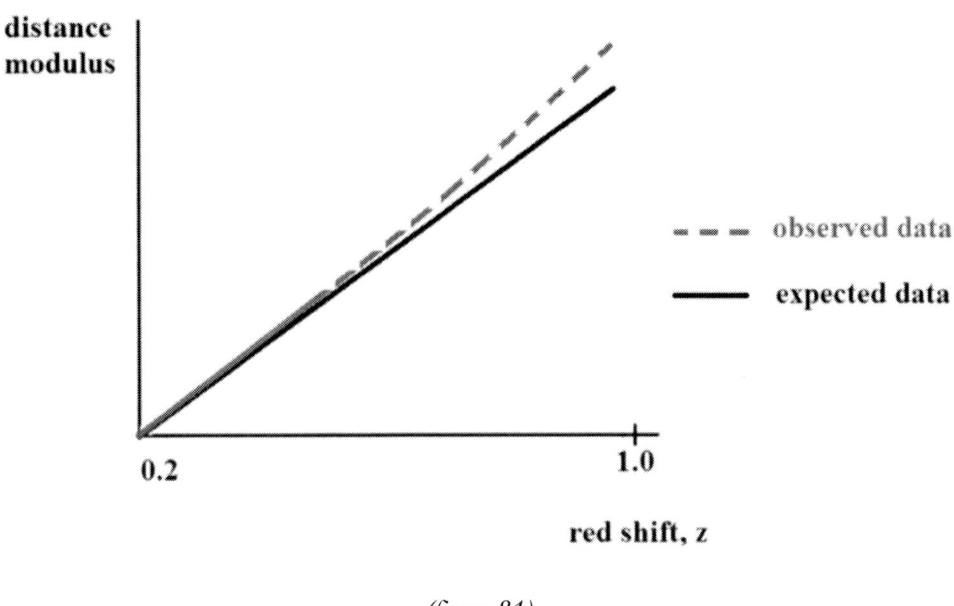

(figure 81)

Dark Energy Candidates

There clearly must be something driving the accelerating expansion of the Universe.

Cosmologists call this *dark energy.*

Whether it is negative or positive energy, that is whether it is attractive or repulsive, it is still an energy component of our Universe.

The actual amount of baryonic matter in the whole Universe is a mere 4%!

The energy spread in the Universe can be summarized as follows .....

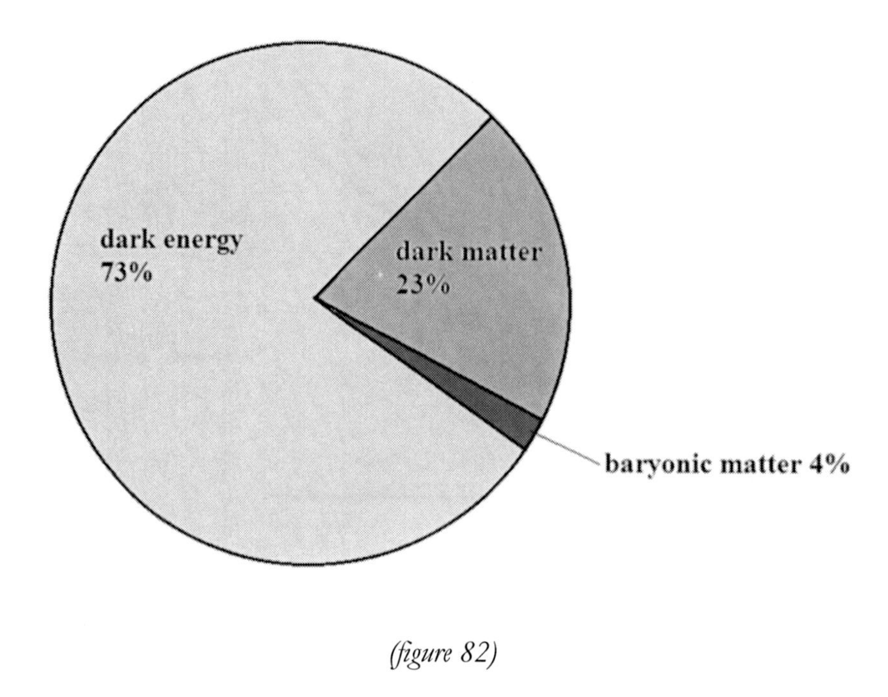

(figure 82)

So what could the candidates be for the source of dark energy?

Einstein's Cosmological Constant

One way of expressing Einstein's field equations which describe spacetime is as follows:

$$G^{\alpha\beta} = 8\pi T^{\alpha\beta}$$

The left hand side of this equation is the *metric tensor.*

This describes the *curvature of spacetime.*

The right hand side of the equation is the *stress-energy tensor.*

This describes the matter/energy content of spacetime.

But Einstein believed that the Universe was not expanding, so he added a term to the equation which is a constant enabling the expansion to be either negative or positive, thus:

$$G^{\alpha\beta} + \lambda g^{\alpha\beta} = 8\pi T^{\alpha\beta}$$

Now, λ is the cosmological constant, and the 'correct' value would render any expansion of the Universe equal to zero.

However all that changed when, by 1929, it was known that the Universe was actually expanding!

But now that constant has found itself in favour again, due to the discovery that the expansion of the Universe is actually accelerating.

Quintessence

Quintessence is a postulated *scalar energy field* which has the property of pushing the cosmos apart.

Its equation of state is defined as the ratio of its pressure to its density.

For quintessence to work, the pressure will be negative.

If the energy density is ρ, then the strength of the gravitational force in the vacuum is given by:

$$\textbf{Gravitational force} = \rho + 3P$$

where **P** is the pressure in the vacuum.

In nearly all cases, the pressure **P** is very small compared with the energy density and the resulting gravitational force is attractive.

However, when the right hand side of the equation is negative, then the gravitational force becomes repulsive.

For quintessence to work, then the pressure must be negative enough to overcome the total attractive gravitational force of the energy density of the Universe.

Question

Data:

Universal Hubble Constant, H = 72 kms^{-1} per megaparsec

1 megaparsec = 3.26 light years

Speed of light in vacuo, c = 3.0×10^8 ms^{-1}

Given that the Hubble law can be expressed by the following …..

$$v = H\,d$$

and that the red shift,

$$z = \frac{\Delta\lambda}{\lambda} = \frac{v}{c}$$

(where 'v' is the recessional velocity of a galaxy,

'd' is the distance from the Earth (in megaparsecs),

And 'H' is the Hubble constant for the Universe)

Calculate the distance away of a galaxy for a red shift of 0.9 in light years.

CHAPTER 18: What If?.......

This last chapter brings us back again to particle physics.

There are some ideas in existence about the *opposite* type of particle that we know about.

We will look at the quite reasonable premise that might inspire those ideas, and then we will carefully examine whether or not it looks feasible!

What if? ………..

Field Symmetry

Electric Fields

Imagine a spherical conductor which is *positively* charged.

The imagined electric field lines would *flow outwards* from the conductor radially and equispaced in all directions, like this…

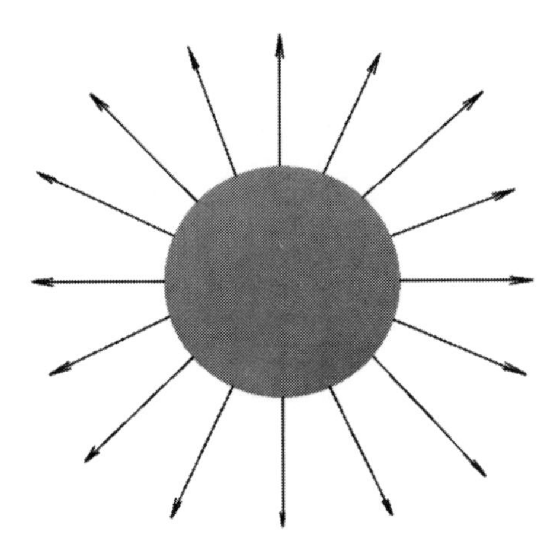

Positively charged sphere

(figure 83)

Suppose now that we had a similar spherical conductor which was *negatively* charged.

In this case the imagined electric field lines would *flow inwards* like this…

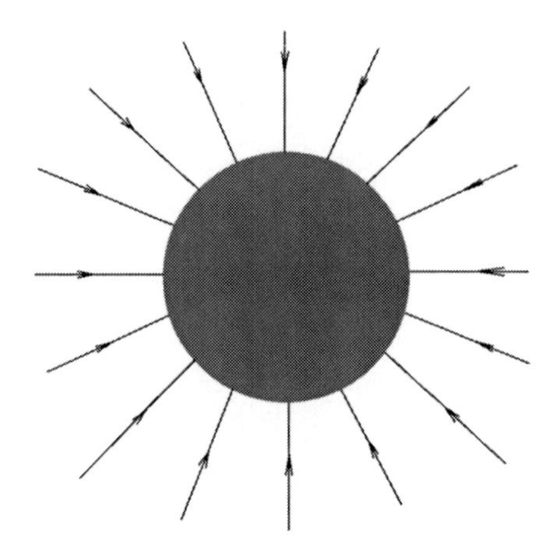

Negatively charged sphere

(figure 84)

Magnetic Fields

Of course, an electron would be repelled from the negatively charged sphere, and attracted towards the positively charged one.

In addition to that, some gauge theories predict that in the early Universe, *magnetic monopoles* would have been created, and should still be observed, although none so far have been found.

This means that a *north monopole* would have its imaginary magnetic field lines flowing *outwards*, in a similar fashion to the negatively charged sphere, like this…

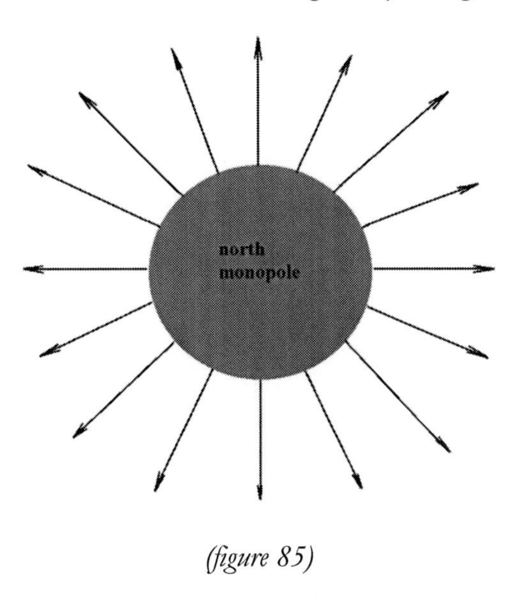

(figure 85)

And similarly, a *south monopole* would have its imaginary magnetic field lines flowing *inwards* like this…

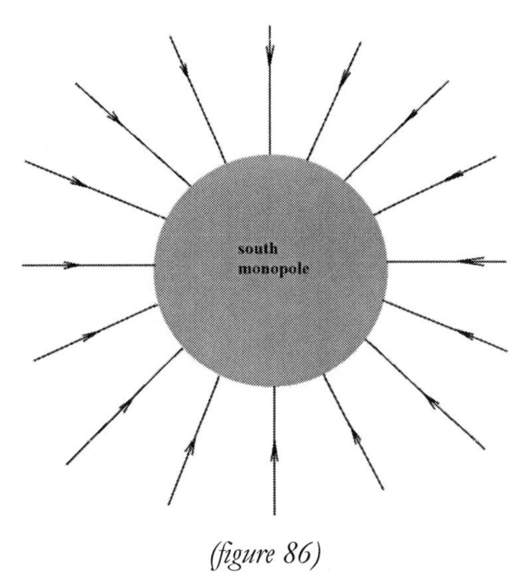

(figure 86)

Gravity Fields

What about gravity fields?

Well we live in a *gravity potential well.*

This means that, just like electric and magnetic fields, imaginary gravity field lines flow *into* all bodies with mass.

For example, if we imagined the Earth or any celestial body to be a solid mass, the field lines would flow *inwards*, like this…

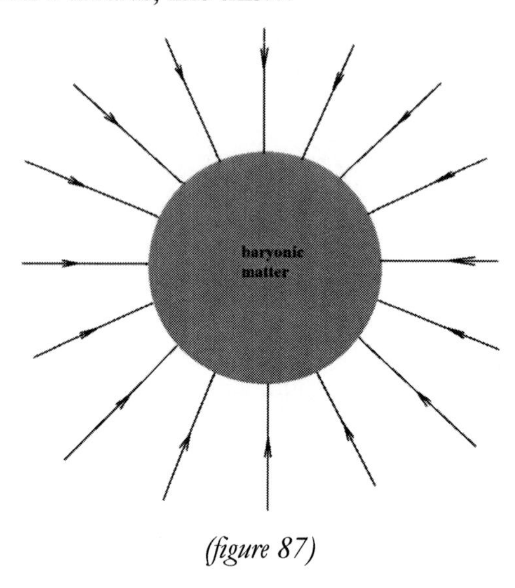

(figure 87)

Now let us look at our fields together…

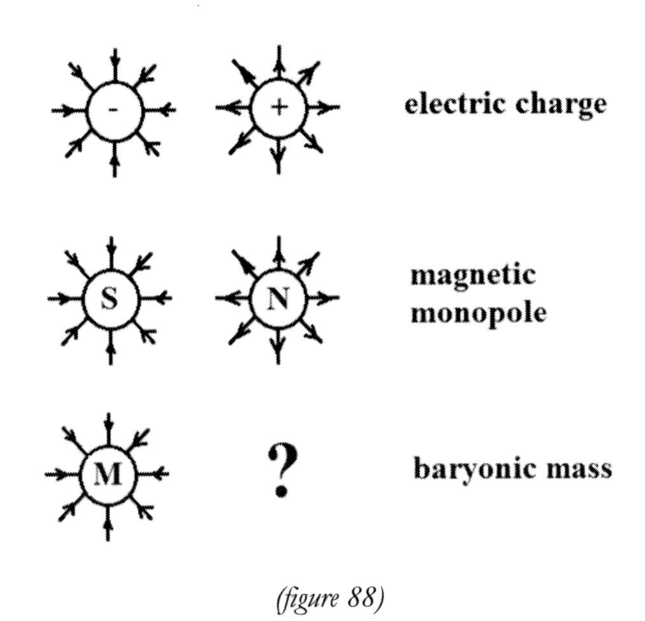

(figure 88)

Does there seem to be something missing?

The answer is – maybe!

Going by the eloquence of symmetry, it does seem that there could be a type of mass *out of which* our imaginary gravity field lines would flow!!!

As yet, no such mass has been observed.

Now let's consider the implications of such a thing.

First of all, all the baryonic matter that we know about is *gravitationally negative.*

So if there did exist a type of matter out of which the field lines would flow, it would be *gravitationally positive.*

That is to say that it would *gravitationally repel* all other matter.

Now there is an obvious anomaly here.

That is that in the case of electric and magnetic fields, it is the *opposite* type of field that attracts, and the *like* field that repels.

However, in the case of baryonic mass, the *like* field attracts, so that all massive bodies, each with their gravitational field lines flowing inwards, attract each other.

Nevertheless, a type of matter or particle *out from which* the gravity field lines extend might exhibit the property of gravitational repulsion.

Let's call such a particle *gravitationally positive.*

So on the face of it, we can't really fault the symmetry presumption which inspires it.

But perhaps when we examine the reality of it we will come to a different conclusion.

Dark Matter and Dark Energy

Imagine if such gravitationally positive matter particles existed.

They would seek to migrate away from all gravitationally negative matter and, apart from some that might be trapped inside galaxies (such that there would be an equal repulsion in *interstellar* space), the particles would tend to congregate in intergalactic space.

What effect might this have?

It could be the solution to the mystery of dark matter – because such a collection of particles would tend to want to push spiral galaxies from the outside.

And that would be something like this…

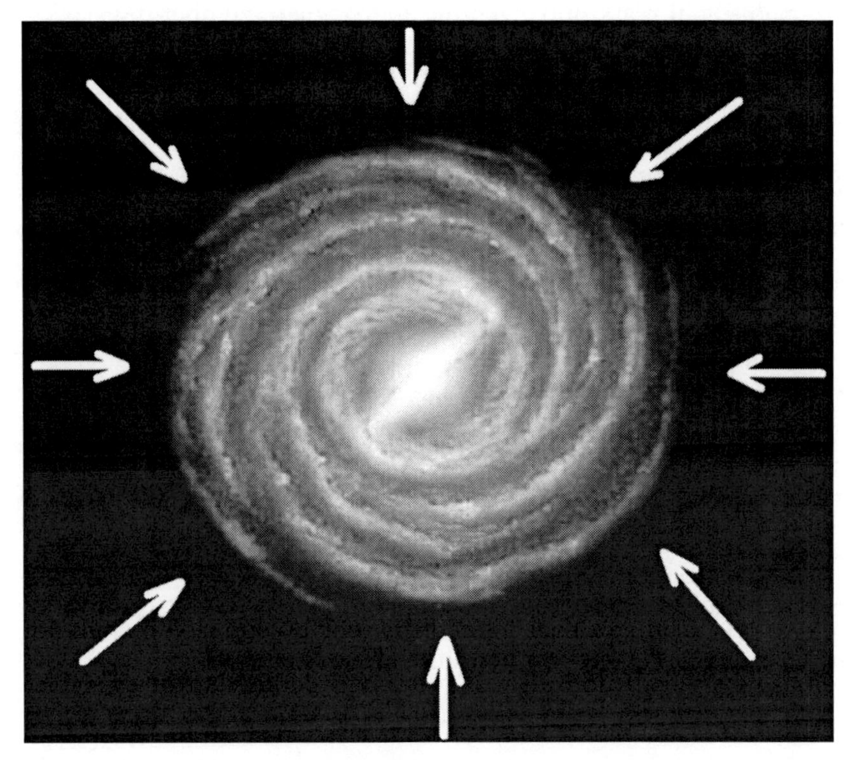

(Courtesy of NASA/JPL-Caltech/R Hurt/SSC)
(figure 89)

Can you think of just one obvious attribute of such a particle?

It would have to be neutral, because if it were not, then electrostatic forces would overwhelm and the gravitational force would be rendered obsolete.

So what sort of neutral gravitationally positive particle would we be considering?

Let's imagine that there exist gravitationally positive 'neutrons'!

Can you work out the approximate number of protons in the visible Universe?

It is in the order of 6.0×10^{78}, and you will have your chance to confirm these calculations soon.

A simple calculation will show that the volume of the visible universe is some 1.08×10^{31} cubic light years, and given the approximate number of galaxies, we can easily calculate the 'volume' of space time, as an average, for each galaxy.

If we imagine the number of gravitationally positive 'neutrons' to equal the number of protons in the Universe, we can then calculate the total repulsive gravitational force exerted by all the gravitationally positive 'neutrons' surrounding each galaxy, spiral galaxies included, assuming that the Universal Gravitational Constant, 'G' is the same for both types of mass.

This comes to around some 8500 N per galaxy.

We can again easily show that the negative gravitational force needed to keep the outliers orbiting a typical spiral galaxy containing 1×10^{11} stars, is some 8.5×10^{20} N.

This shows that the force from our proposed gravitationally positive 'neutrons' is lacking by a factor of some one hundred billion Newtons!

The only way that this could be rectified is to assume that there are some one hundred billion more gravitationally positive 'neutrons' in the Universe than there are protons.

This seems completely ridiculous.

Thus our conclusion must be that the proposal is very probably nonsense!

Further, if such particles can't rise to the job of keeping all the matter in spiral galaxies rotating, there seems little chance that they can also be responsible for accelerating the expansion of those galaxies!

It is worth also noting that there is a proposal in some quarters that *antimatter* itself is gravitationally positive.

Admittedly, we don't know for sure if this is true or not, but there are many very good arguments to indicate that this is not the case.

Could this be tested?

In theory it could.

How could it be done?

In a long linac in which an antimatter beam could be projected, we could in theory measure if there were any uplift in the particle's trajectory in the course of its travel down the linac.

If it was gravitationally positive, then it would be repelled by the Earth's negative gravitation.

But could such a deflection be measured?

In the examples which follow you can perform just such a calculation and you can make your own conclusion!

Conclusion

It doesn't take much effort to see that the premise is sadly deficient.

With relief we stay with the standard model of particle physics, and hope that soon we will discover the true nature of the dark matter and dark energy in our Universe!

We look to the LHC at CERN for breakthroughs in this field in the not too distant future.

Questions

1.

Data:

Universal Gravity Constant, G = 6.673×10^{-11} N m^2 kg^{-2}
Speed of light in vacuo = 3.0×10^8 ms^{-1}
Average mass of galactic star is 7.1963×10^{26} tonnes
Number of stars in average galaxy = 1×10^{11}
Number of galaxies in Universe = 1×10^{11}
Radius of average galaxy = 50,000 light years
Rest mass of proton = 1.673×10^{-27} kg
Take rest mass of gravitationally positive 'neutron' to be 1.675×10^{-27} kg
Radius of visible Universe = 13.7×10^9 light years

(a) Calculate the number of protons in an average star, assuming total composition of hydrogen

(b) Hence calculate the number of protons in the visible Universe

(c) Calculate the volume of the Universe in cubic light years

(d) Assuming equispacing of galaxies, calculate the total volume occupied by each galaxy

(e) Assuming the number of hypothetical gravitationally positive 'neutrons' to equal the number of protons in the Universe, calculate the number of such particles available which might surround each galaxy

(f) Assuming equispacing of such particles throughout each galactic space, calculate the separation of each particle

(g) Hence calculate the total available compressive force from such particles

(h) Assuming the amount of dark matter in each galaxy is some seven times the mass of the galaxy, calculate the additional force to the baryonic matter gravitational attraction needed to keep all the stars in a spiral galaxy rotating

(i) Hence calculate the deficit

2.

Data: **Assume speed of positron in linac to be 0.98c**

Take the speed of light in vacuo, 'c', to be 3.0×10^8 ms^{-1}

Rest Mass of positron = 9.11×10^{-31} kg N m^2 kg^{-1}

Mass of Earth = 5.978×10^{24} kg

Radius of Earth = 6.378×10^6 m

A three kilometre linac is used to project the antimatter particle to an electron (a positron) along its length.

Neglecting relativistic effects:

(a) Calculate the amount of time it takes the positron to cover one kilometre of linac track

(b) Calculate the upwards force on the positron

(c) Calculate the upward acceleration of the positron

(d) Assuming the positron is gravitationally positive; calculate the amount of elevation that it will achieve after traversing one kilometre of track

(e) Calculate the amount of elevation it would achieve after traversing three kilometres of track

(f) Is it a realistic prospect to be able to measure this distance?

CHAPTER 19: The Large Hadron Collider

The LHC

The Large Hadron Collider is situated in both Switzerland and France and is operated by CERN (Organisation Européenne pour la Recherche Nucléare).

What does 'Large' Hadron Collider mean?

Do we mean that the collider is very large?

Or do we mean that it will collide 'large' hadrons, because if we consider 'large' hadrons to be related to the number of quarks, then it is – it will collide protons, which are baryons and contain three quarks (and also lead ions) as opposed to mesons which contain just one quark and an antiquark.

However, if we consider mass then some mesons are more massive than some baryons!

We will clearly refer to 'large' as to the physical size of the synchrotron.

It is some 27 km in circumference and contains some 9300 magnets!

(Reproduced by kind permission of CERN – CERN-MI-0807031 Photograph: Maximilien Brice ©CERN Geneva)
(figure 90)

The history of CERN goes back as far as 1954 when member nations in Europe founded the organisation.

The first accelerator came into commission in 1957 when the 600MeV synchrocyclotron started operation. *(19.1)*

Perhaps CERN's history is best known for the LEP (Large Electron Positron) collider which started operation in 1989.

Here electrons were collided with their antiparticle, the positron, at 100GeV in the same 27 km tunnel that is currently being used. *(19.2)*

The LEP was and still is the largest electron positron collider that the world has ever known.

But as far as its *present day* operation is concerned, it is also famous for something else – unless some other species in some other star system has built one - it is the *coldest place* in the Universe!

The magnets operate at just 1.9K, which is colder than intergalactic space which the CMBR tells us is just 2.725K! *(19.3)*

When the collider is operating on full power, two proton beams will traverse the circuit, each with a total energy of 7TeV (7 x 10^{12} eV), which will result in head on collisions with an energy of 14TeV! *(19.4)*

There could be up to some 600 million proton collisions per second, and the analysis of such a cohort requires some of the most advanced detection and analysis components in the world.

In terms of speed, protons at full power will *enter* the main synchrotron ring at "0.999997828c, and will then be accelerated up to 0.999999991c at 7TeV per beam". *(19.5)*

In order to achieve this, the beams will be accelerated around the synchrotron for some 20 minutes *(19.6)*

Given that the protons will complete 11,245 circuits of the synchrotron per second *(19.7)*, calculate the approximate number of circuits that each proton will make in order to achieve full power.

What did you get?

It comes out at some thirteen million, four hundred and ninety four circuits!

Think now about how difficult it is to accelerate something with relativistic mass up to very high relativistic speeds!

At the time of writing, CERN is operated by "twenty European Member States, with scientists from some five hundred and eighty world-wide institutions and eighty five nationalities using its facilities"! *(19.8)*

However, do not assume that every scientist working at CERN is looking for the same thing - a former student of the author is working there, looking for the Higgs Boson.

However, this physicist does not expect to find it and is a sceptic!

The Experiments

There are six experiments at CERN

These are ATLAS, CMS, ALICE, LHCb, TOTEM, and LHCf.

ATLAS

This is '**A T**oroidal **L**arge hadron collider **A**pparatu**S**'

Its purpose is to search for the Higgs boson, and to possibly look for the existence of other dimensions.

The gravitational force is extremely weak in comparison with the other forces in nature, and it has been postulated that this could be because gravity is the only force that is shared amongst other dimensions.

Evidence for other dimensions could possibly be seen if a particle under observation suddenly disappeared without a trace, or a new particle suddenly appeared without the energy available for its creation.

Another possibility is that ATLAS will find evidence for dark matter particles.

"The ATLAS detector is 46m long, 25m high and 25m wide and weighs in at some 7000 tonnes".

This makes it the largest volume detector ever built *(19.9)*

Compare the size of the detector to that of the person standing nearby…

(Reproduced by kind permission of CERN – CERN-EX-0702041 Photograph: Maximilien Brice; Claudia Marcelloni ©CERN Geneva)
(figure 91)

CMS

This is a '**C**ompact **M**uon **S**olenoid' detector.

The CMS detector is a general purpose detector as is the ATLAS detector.

It will also look for evidence of the Higgs boson, dark matter particles, and extra dimensions, but it uses different technology.

It is "21m long, 15m wide and 15m high and it contains a huge solenoid magnet which can generate a magnetic field some 100,000 times stronger than that of the Earth (about 4T) and it weighs in at some 12,500 tonnes". *(19.10)*

Again, compare its size of part of it with that of the people shown

(Reproduced by kind permission of CERN – CMS-PHO-MUON-2007-001-4 Photograph: Michael Hoch ©CERN Geneva)
(figure 92)

ALICE

This is '**A** **L**arge **I**on **C**ollider **E**xperiment'.

This outstanding experiment will collide *lead ions* at very high energies to try and simulate conditions just after the Big Bang occurred.

The energies and temperatures achieved are such that it is expected that the quarks in the nuclei will actually be freed and will be able to exist outside of their strong force bonds, with the aim of creating a quark gluon plasma, just as in the very early moments after the Big Bang.

By studying the behaviour of the plasma, a better understanding is expected of how the particles we know about were formed.

The ALICE detector weighs in at some "10,000 tonnes, is 26m long, 16m high and 16m wide". *(19.11)*

The following outstanding and historic results were achieved:

On 7th November 2010, a temperature of 1×10^{13} K was achieved in this detector.

This is about 800,000 times the temperature in the core of our Sun and a quark gluon plasma was, for the first time, observed. *(19.12)*

Once again, compare its size…

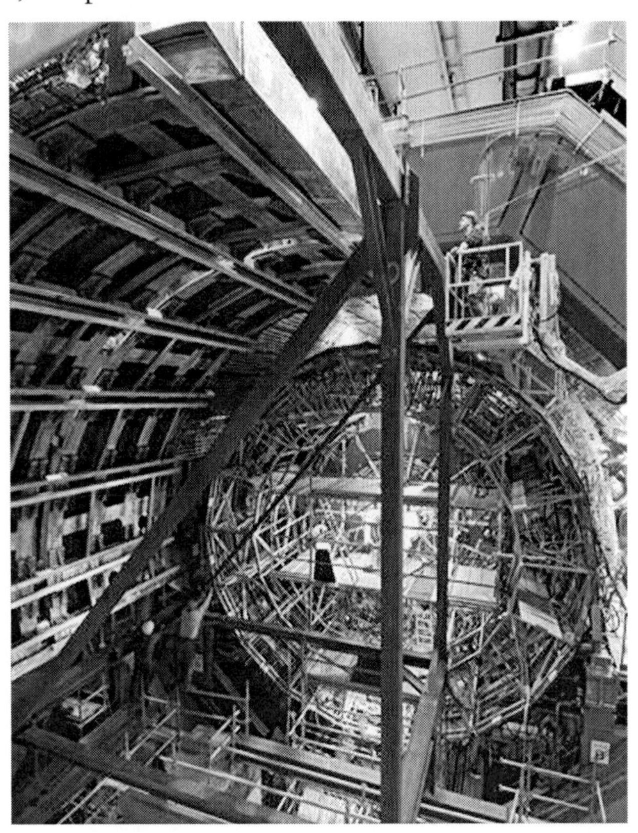

(Reproduced by kind permission of CERN – CERN-EX-0711012-05 Photograph: Mona Schweizer ©CERN Geneva)
(figure 93)

LHCb

This is '**L**arge **H**adron **C**ollider **B**eauty'.

Why is it called 'beauty'?

It isn't because the detector is any more beautiful than any of the others, but it does refer to the predominant type of quark that it will be concerned with.

This is the 'beauty' quark, otherwise known as the 'b' quark or bottom quark.

In some circles, bottom quarks are still referred to as beauty quarks.

What is the main purpose of the experiment?

This is to try to investigate why the Universe seems to be made from matter.

Perhaps the great mystery of where all the antimatter went, or how matter predominated over antimatter in the Big Bang, will be solved.

Bottom quarks and antibottom quarks existed in great numbers in the early stages of the Universe just after the Big Bang, and this experiment will create them again in large numbers.

What will be looked for?

An imbalance – any imbalance in the number of matter particles to antimatter particles will advance our understanding of what might have happened in the Big Bang which gave us the strange matter dominated Universe we occupy.

And remember – one of our biggest mysteries still is – where did all the antimatter go?

The explanations that we have so far for an imbalance, such as CP (charge parity) violation would have resulted in a far too small excess of matter over antimatter in the early Universe, so we still are faced with a big mystery.

This detector weighs in at "5600 tonnes, is 21m long, 10m high, and is 13m wide".

Here is a cutaway section of the detector…

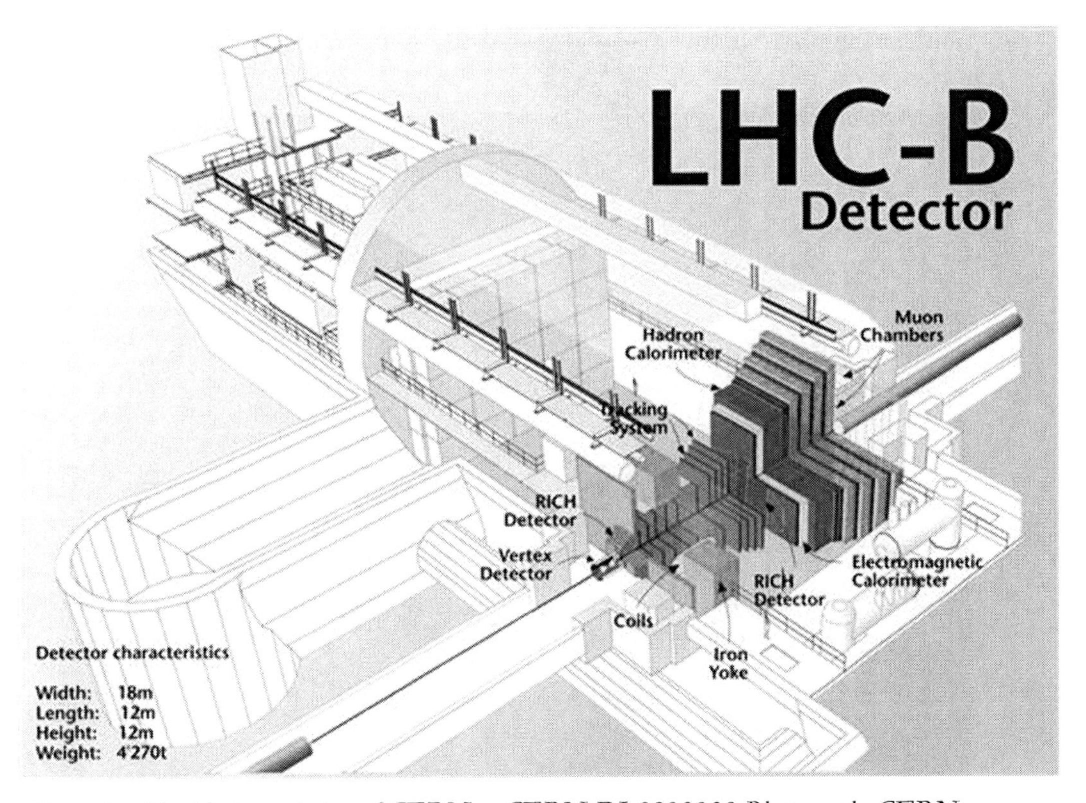

(Reproduced by kind permission of CERN – CERN-DI-9803030 Photograph: CERN AC ©CERN Geneva)
(figure 94)

Other Results

Has the LHCb experiment shown us anything else?

It certainly has.

It has given us excellent evidence of the decay of a **Z** boson.

Look at the photo below…

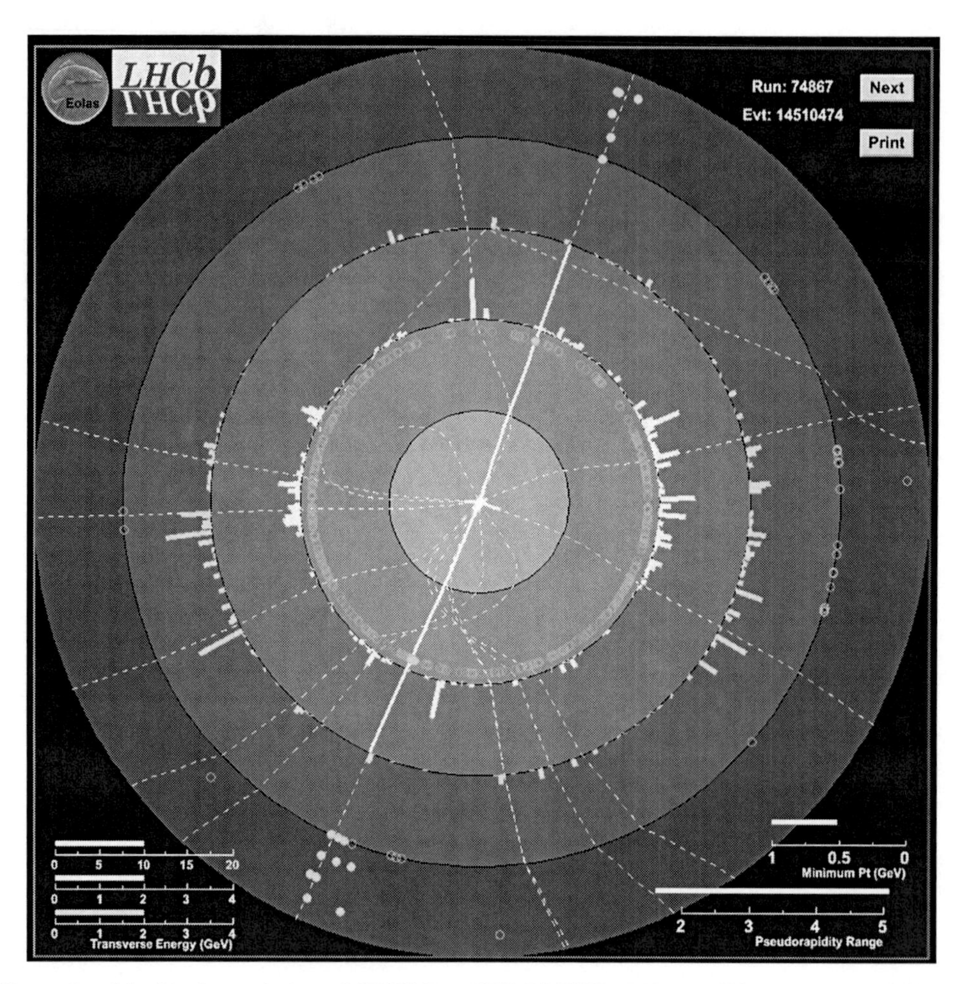

(Reproduced by kind permission of CERN – CERN-EX-1007138 Photograph: LHCb Collaboration ©CERN Geneva)
(figure 95)

"The **Z** boson has decayed immediately into two muons, indicated by the two thick white lines which result in the circular event results in the muon chamber at the end of the thick white lines" *(19.13)*

In addition, the LHCb detector has caught first rate evidence of a **W** boson in action.

Look at the following picture…

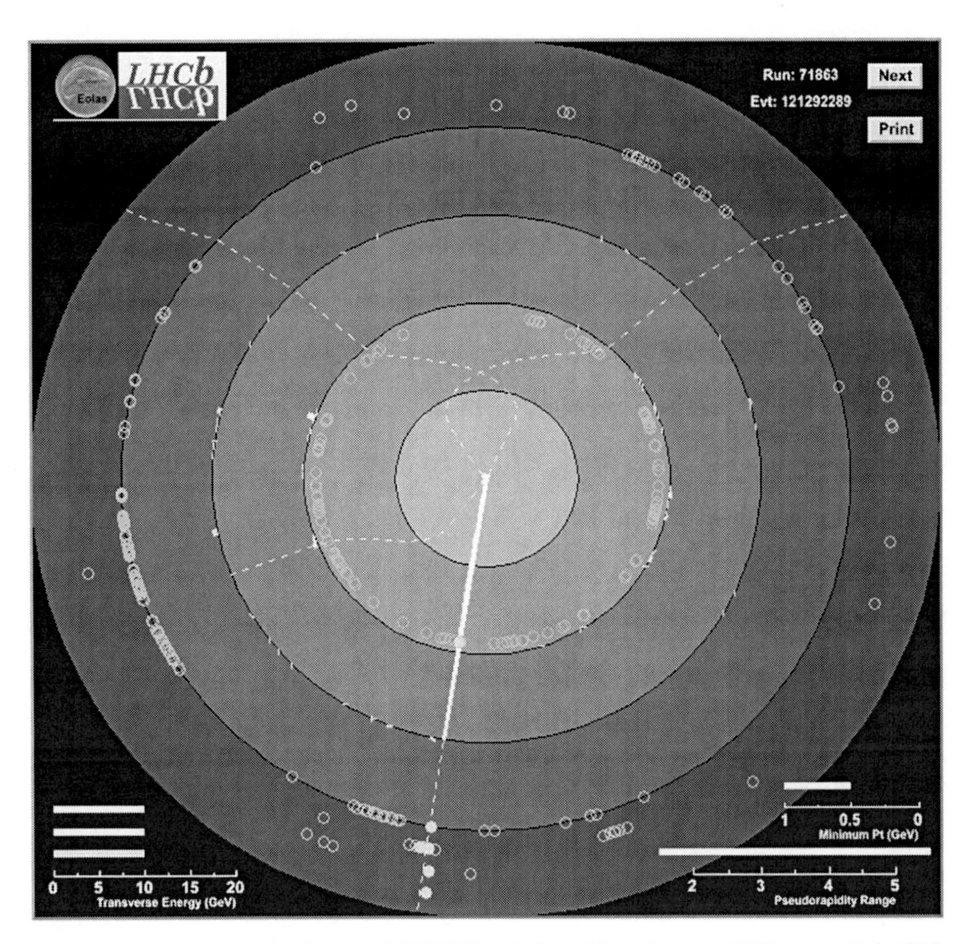

(Reproduced by kind permission of CERN – lhcb-public.web.cern.ch Photograph: LHCb Collaboration © CERN Geneva)
(figure 96)

The **W** boson, as we know, is the weak force boson, which results in radioactive decay when it operates.

In the above picture, the boson "decays at the centre and immediately decayed into a muon, which is shown by the thick white line which in turn is directed at the circular results in the muon detector at the bottom of the picture". *(19.14)*

There remain now only two much smaller experiments in the Large Hadron Collider:

TOTEM

This stands for '**TOT**al **E**lastic and diffractive cross section **M**easurement'.

This experiment complements the CMS experiment and is much smaller, weighing in at just 20 tonnes.

It studies the "precise proton proton interaction cross section measurement, but in addition to that it will try to shed more light on the actual proton structure and measure the size of the proton". *(19.15)*

LHCf

This stands for '**L**arge **H**adron **C**ollider **F**orward'.

This is the smallest experiment, with just two 40 kg detectors.

However, its size bears no reflection on how interesting and important its findings are.

Its aim is to shed more light on exactly what is happening when cosmic rays from space interact with atoms in the upper Earth's atmosphere.

Let's remind ourselves of some of the particles which might be observed resulting from the interaction of cosmic rays with nuclei in the Earth's upper atmosphere.

These will include some decay products as well.

There could be showers of pi mesons, muons, electrons, positrons, neutrinos, neutrons and protons, and gamma ray photons amongst other possibles.

Sometimes millions of cosmic ray particles will strike the Earth's upper atmosphere in a heavy bombardment.

By studying how collisions in the experiment can produce similar arrays and showers of particles will assist in the interpretation and calibration of large scale cosmic ray experiments that can be conducted. *(19.16)*

Question

Data:

Speed of light in vacuo = 2.997924591 x 10^8 ms^{-1}

Electronic charge = 1.6021927 x 10^{-19} C

Rest mass of proton (kg) = 1.67261411 x 10^{-27} kg

Rest mass of proton (MeV) = 938.2595 MeV

When any massive body is travelling at relativistic speeds, its mass will be greatly increased.

(a) Calculate the relativistic mass (in kg) of a proton travelling in a synchrotron at a speed of 0.9999999c

Note that this speed is considerably less than the full power speed of a proton in the CERN synchrotron.

(b) How many times greater is this than the rest mass of a proton?

(c) It is much more usual to perform calculations about an object with relativistic speed in terms of energy rather than mass.

Calculate the 'mass' of the relativistic proton now in terms of energy (in TeV)

(d) Calculate the relativistic mass of a 1 tonne spacecraft travelling at 0.9999999c.

(e) Is it a realistic project to accelerate such a space probe to this speed?

(f) What value of relativistic mass does a body with rest mass tend towards as its velocity approaches light speed?

Use the following mass and energy relativistic equations:

To find the relativistic mass, m:

$$m = \frac{m_0}{\sqrt{1 - \frac{v^2}{c^2}}}$$

Where m_0 is the rest mass

v is the relativistic speed

To find the relativistic energy equivalent, E:

$$E = \frac{m_0 c^2}{\sqrt{1 - \frac{v^2}{c^2}}}$$

Where m_0 is the rest mass

v is the relativistic speed

CHAPTER 20: Answers and Solutions

Chapter 2

(a) No

(b) The unstable particle 'disappears from the Universe' only in the sense that the particle degenerates into energy.

This is possible due to the equivalence of mass and energy, shown by Einstein's famous equation.

New more stable particles are then produced from the energy

Chapter 3

1.

(a)

$\mathbf{E = mc^2}$

Hence, $E = (9.10957 \times 10^{-31}) \times (2.997924 \times 10^{8})^{2}$

So $E = 8.18727 \times 10^{-14}$ joule

So $E = \dfrac{8.18727 \times 10^{-14}}{1.60219 \times 10^{-19}}$

So $E = 5.11 \times 10^{5}$ eV

So $E = 0.511$ MeV

(b)

$E = 939.55$ MeV

Chapter 4

(a) 16 MeV

(b) 938.2595 MeV

(c) 1.7 %

(d) 98.3 %

(e) Refer chapter 10 '1. Strong Nuclear Force' just before '2. Electromagnetic Force'

Chapter 5

1. (K^-, K^+) and (π^-, π^+). The Δ^- baryon is not the antiparticle of the Δ^+ baryon (although it may look like it from the mass and lifetime) as you will see from the quark structure in the next chapter. However antimesons can switch between quark and antiquark and hence switch between meson and antimeson, again as you will see from the quark structure in the next chapter.

2. (a) ⅓ (b) 0 (c) -⅓ (d) 1 (e) 0 (f) ⅓
   (g) 1 (h) -1 (i) 0

3.

(a) all four conservation laws balance

(b) mass imbalance so not permitted

(c) charge imbalance so not permitted

(d) all four conservation laws balance

Chapter 6

1. $n + \Sigma^0 \rightarrow p + (X)$

   Charge: $(0) + (0) \rightarrow (+1) + (X)$

   Hence X must have a charge of -1

   Lepton No: $(0) + (0) \rightarrow (0) + (X)$

   Hence X is not a lepton

   Baryon No: $(1) + (1) \rightarrow (1) + (X)$

   Hence X is a negative baryon

   Quark balance: (u d d) + (u d s) $\rightarrow$ (u u d) + (X)

   Hence X must be (d d s)

   So X is a $\mathbf{\Sigma^-}$ baryon

2. X is $(d\ \bar{s})$ which is a $\mathbf{K^0}$ meson

3. X is $(d\ \bar{u})$ which is a $\boldsymbol{\pi}^-$ meson

4. X is $(c\ \bar{d})$ which is a $\mathbf{D^+}$ meson

Chapter 7

$$\gamma + p \rightarrow x + \bar{D}^0 + n$$

Charge conservation: $(0) + (+1) \rightarrow (\mathbf{x}) + (0) + (0)$

Hence, **x is positive**

Baryon number: $(0) + (1) \rightarrow (\mathbf{x}) + (0) + (1)$

Hence, **x is not a baryon**

There are no leptons in the equation

Hence, **x is a positive meson**

Quark structure: $(0) + (uud) \rightarrow (\mathbf{x}) + (u\ \bar{c}) + (udd)$

Hence, **x must have quark structure $(c\ \bar{d})$**

Hence, **x must be a D^+ meson**

Chapter 8

The charge on a particle passing through a bubble chamber is primarily responsible for the ionisation.

It is the charge that will induce electrons to become ionised from the hydrogen atoms resulting in the ionisation

Chapter 9

1. Although these two tracks show the presence of charge particles, due to their circular arcs, and they both have the same energy, due to their equal radius, these two tracks do not show a matter antimatter pair because they both depart the creation point in the same direction, that is, they both describe clockwise circular arcs from the interaction point!

2.

(a) the point 'X' is approached by a neutral particle

(b) at point 'X' either annihilation occurs and new particles are created or the neutral particle decays and new particles are created

(c) two of the particles are negatively charged and two are positively charged

(d) two matter antimatter pairs could possibly have been created but one pair might have different energies

(e) If the two outermost tracks indicate a particle antiparticle pair the one with the greatest energy is the lower one to the right since it has a larger radius of curvature

3.

(a) A third unknown particle, 'Z', is created there

(b) The particle propagates and reveals itself as neutral since no track is evident

(c) The neutral particle decays into a negative pi meson (π^-) and a positive pi meson (π^+)

(d) These are a matter antimatter pair

(e) The equation representing the event is

$$\mathbf{K^- + p \rightarrow p + \pi^- + Z}$$

Charge, Q: $(-1) + (+1) \rightarrow (+1) + (-1) + (\mathbf{Z})$

Hence the particle is neutral

Lepton No L: $(0) + (0) \rightarrow (0) + (0) + (\mathbf{Z})$

Hence the particle is not a lepton

Baryon No B: $(0) + (1) \rightarrow (1) + (0) + (\mathbf{Z})$

Hence the particle must be a neutral meson

Quark Structure: $(s\ \bar{u}) + (u\ u\ d) \rightarrow (u\ u\ d) + (d\ \bar{u}) + (\mathbf{Z})$

Hence **Z** must be $(s\ \bar{d})$

This means that the unknown particle, **Z**, is a neutral K antimeson ($\bar{K}^0$):

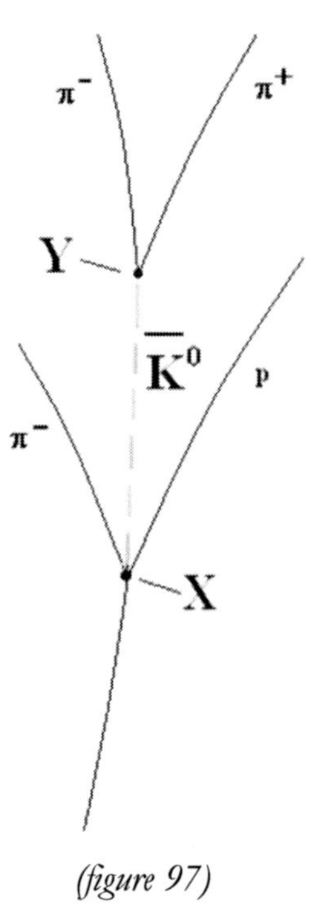

(figure 97)

Note that during its propagation from point X to point Y the K^0 antimeson will oscillate between the meson/antimeson states of K^0 and $\bar{K}^0$. This phenomenon is called *particle oscillation.*

Chapter 10

1. (a) 1000.95 N

   (b) probably not considering small wave motion and wind factors

2. 4.165 x 10^{42} times greater. This applies at all separations

Chapter 11

The masses of the weak force bosons, the only gauge bosons to have mass, is achieved by examining the *energies and masses* of the decay products of the bosons.

In the case of the **W** bosons, the masses and energies of the resulting electrons and electron neutrinos are carefully monitored.

The total mass/energy resulting from the decay gives an approximation of the boson mass.

Bear in mind that many measurements must be carried out in order to achieve an acceptable precision

For example, in the case of the Fermi National Laboratory DZero experiment in 2009, in the U.S.A., a mass precision was achieved of 0.05 percent, giving a **W** boson mass of 80.401 +/- 0.044 GeV [ref.(20.1)]

Chapter 12

1.

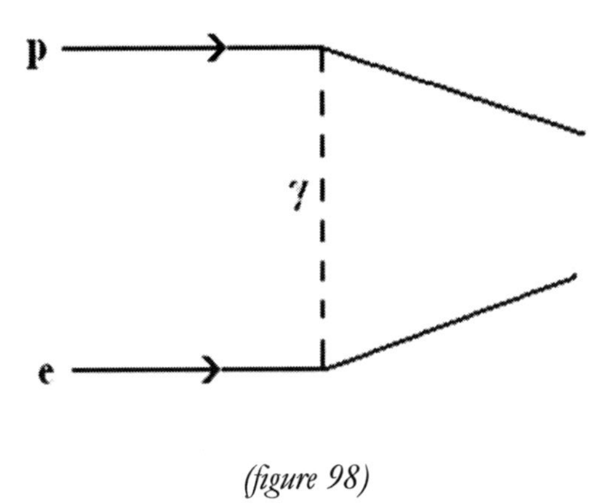

(figure 98)

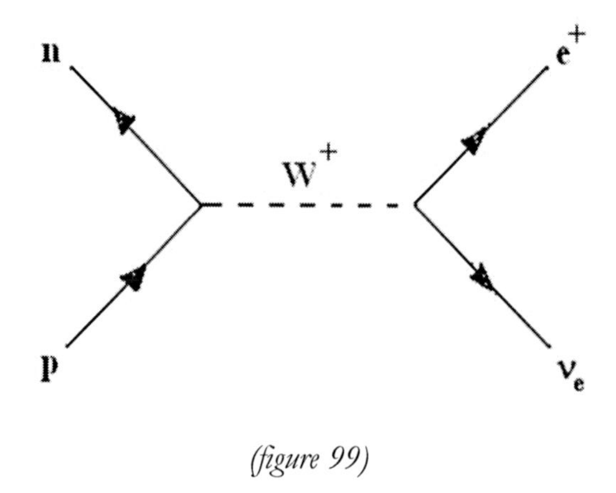

(figure 99)

3. A positive K meson, **K$^+$,** (u $\bar{s}$), exists at the start, in which the strange antiquark decays by the mediation of a weak **W**$^+$ boson into an up quark and down antiquark. This results in the production of two pi mesons, a positive pi meson, π^+, (u $\bar{d}$) and a neutral pi meson, π^0 (u $\bar{u}$).

Chapter 14

1.

4.775×10^{52} years or nearly five hundred million billion billion billion billion billion years.

2.

(a) 1×10^{-30} metre

(b) 100

(c) 1.2677×10^{30} times or one thousand billion billion billion times

(d) About 1.2677 m

Chapter 16

(a) 4.7304×10^{20} m

(b) 1.43926×10^{41} kg

(c) 1.4249×10^{5} ms^{-1}

(d) 3.7699×10^{5} ms^{-}

(e) 2.6457 times

(f) actual mass = 1.0074799×10^{42} kg

(g) seven times

(h) seven times

Chapter 17

Since $\mathbf{v = Hd}$

And $\mathbf{v = zc}$

Then $$\mathbf{d = \frac{zc}{H}}$$

$$\mathbf{d = \frac{0.9 \times (3.0 \times 10^5)}{72}}$$

Hence, d = 3750 megaparsecs

So, d = 12.225×10^{9} light years

Chapter 18

1.

(a) 4.3014345×10^{56}

(b) 4.3014345×10^{78}

(c) 1.08×10^{31} (light years)^3

(d) 1.08×10^{20} (light years)^3

(e) 6.0×10^{67} particles

(f) 1.151 m

(g) 8480 N

(h) 8.35×10^{20} N

(i) approximately 8.35×10^{20} N!

2.

(a) 3.4×10^{-6} s

(b) $8.9335975 \times 10^{-33}$ N

(c) 9.81×10^{-3} ms^{-2}

(d) 5.67×10^{-14} m

(e) 1.7×10^{-13} m

(f) Since this is less than a quarter of a billion billionth of a metre, it is totally unrealistic!

Chapter 19

(a) $3.7400788 \times 10^{-24}$ kg

(b) 2236.068 times

(c) 2.098 TeV

(d) 2.236068×10^{6} kg = 2236.068 tonnes

(e) Probably not

(f) Infinity

References

10.1 Institute of Physics Physics World Vol 23 No 12 Dec 2010 LHC p.5

16.1 Physics World vol 24 no. 1 Jan 2011 p. 4

19.1 http://public.web.cern.ch/public/en/about/History54-en.html

19.2 http://public.web.cern.ch/public/en/about/History89-en.html

19.3 http://public.web.cern.ch/public/en/lhc/Facts-en.html

19.4 http://public.web.cern.ch/public/en/lhc/Facts-en.html

19.5 http://cdsmedia.cern.ch/img/CERN-Brochure-2009-003-Eng.pdf

19.6 http://cdsmedia.cern.ch/img/CERN-Brochure-2009-003-Eng.pdf (p.12)

19.7 http://public.web.cern.ch/public/en/lhc/Facts-en.html

19.8 http://public.web.cern.ch/public/en/About/Global-en.html

19.9 http://public.web.cern.ch/public/en/lhc/ATLAS-en.html

19.10 http://public.web.cern.ch/public/en/lhc/CMS-en.html

19.11 http://public.web.cern.ch/public/en/lhc/ALICE-en.html

19.12 New Scientist 25 Dec.2010 No. 2792/93 p.27

19.13 http://lhcb-public.web.cern.ch/lhcb-public/

19.14 http://lhcb-public.web.cern.ch/lhcb-public/

19.15 http://totem-experiment.web.cern.ch/totem-experiment/

19.16 http://public.web.cern.ch/public/en/lhc/LHCf-en.html

20.1 sciencedaily.com/releases/2009/03/090311153414.htm

Index

Lightning Source UK Ltd.
Milton Keynes UK
UKOW011811071212

203342UK00006B/314/P